# Introduction to Machine Olfaction Devices

# Introduction to Machine Olfaction Devices

Najib Altawell

*Bsc, MRs, PhD, PGCHE, PGCTT*

**ACADEMIC PRESS**

An imprint of Elsevier

Academic Press is an imprint of Elsevier
125 London Wall, London EC2Y 5AS, United Kingdom
525 B Street, Suite 1650, San Diego, CA 92101, United States
50 Hampshire Street, 5th Floor, Cambridge, MA 02139, United States
The Boulevard, Langford Lane, Kidlington, Oxford OX5 1GB, United Kingdom

**Notices**
Knowledge and best practice in this field are constantly changing. As new research and experience broaden our understanding, changes in research methods, professional practices, or medical treatment may become necessary.

Practitioners and researchers must always rely on their own experience and knowledge in evaluating and using any information, methods, compounds, or experiments described herein. In using such information or methods they should be mindful of their own safety and the safety of others, including parties for whom they have a professional responsibility.

To the fullest extent of the law, neither the Publisher nor the authors, contributors, or editors, assume any liability for any injury and/or damage to persons or property as a matter of products liability, negligence or otherwise, or from any use or operation of any methods, products, instructions, or ideas contained in the material herein.

**British Library Cataloguing-in-Publication Data**
A catalogue record for this book is available from the British Library

**Library of Congress Cataloging-in-Publication Data**
A catalog record for this book is available from the Library of Congress

ISBN: 978-0-12-822420-5

For Information on all Academic Press publications
visit our website at https://www.elsevier.com/books-and-journals

*Publisher:* Mara Conner
*Acquisitions Editor:* Carrie Bolger
*Editorial Project Manager:* Rachel Pomery
*Production Project Manager:* Anitha Sivaraj
*Cover Designer:* Greg Harris

Typeset by MPS Limited, Chennai, India

# Dedication

This book is dedicated to all the researchers, educators, engineers, and technicians working within the field of machine olfaction devices across the world.

# Contents

**Contents**

# Contents

**Contents**

Contents

# Contents

# List of Figures

# List of Figures

# List of Figures

# List of Tables

# List of Tables

# List of Boxes

## List of Boxes

# Preface

Universal evolution involves everything, whether in the form of biological evolution (species, including humans) or as simple as the designing of manufactured products, that is, the evolution of life and things—nothing meant to stay the same but everything change.

The process of updating, another term for describing evolution, is happening and progressing all the time, knowing or unknowingly. This process can be noticed clearly years later or by the future generations, looking back to what was in the distant past that differs in various ways compared to the present.

The evolution of the human mind, in principle, follows similar cycles of development. It means our understanding of ourselves and things around us, individually, or part of the society, plays a major role on how our future may develop and shape; in other words, our external environment plays a major role in our evolution. As a part of our learning, humans from ancient times contemplate, worship, or mimic of what nature has provided us. By doing so, it was, and still is, a part of our learning process and consequently a part of our evolution.

As we are born out of this natural environment, our sense of belonging and curiosity does not divert us from what nature provided us with. Our bodies are nothing, but biological machines equipped with what is needed to survive in this earthly environment. Our five senses are vital development evolved to the level we have them at this part of our time. The ability to smell is one of these five senses which plays an important part in our bodily function and safety. However, trying to replicate this function within a machine is another matter, even if the success in this field, which is already happening right now, brings great benefits to our daily lives, the development will continue. This is a basic fact of the human mind and their continuous search for better outcome. Having said that, focusing in this book on this sense and replicating the function within a device is a very small step within a large global scale presently pushing forward within the field of research and development for better outcome. Therefore this book is just another example within a field that many researchers, writers, and developers tread this path long before the idea of the first attempt made many decades ago by trying to design machine olfaction device (MOD).

As an introduction to MOD, this book covers some basic approaches that will enable anyone to follow and understand the materials provided without the need for advanced knowledge of the subject or the need for higher academic qualifications. For this reason the book is written keeping in mind the above aim. Obviously, this does not mean that a student doing a degree or even a researcher within this field may not benefit from it, rather the opposite is true. This is mainly because the simple approach used in explaining and presenting the subject meant everyone can benefit from the materials provided within all the pages of this book.

Finally, the new MOD design provided, as well as the discussion and analysis in various parts of the book, encourage the learner further in this field, irrespective of his/her level or stage in life in pursuing this kind of technology further, whether in the form of new ideas, new design, additional research, or even a new career in this field which in turn may further help in advancing this vital technology for a variety of important needs and aspects in our present and future lives.

**Najib Altawell**

# Acknowledgments

I would like to thank Dr. Michael Cooper for revising the full manuscript. Also, I would like to thank Mr. James Milne for revising Chapter 1, Background, Materials, and Process, and Chapter 2, Comparison and Validations.

# List of Abbreviations

| | |
|---|---|
| **ANi** | Polyaniline |
| **ANN** | Artificial Neural Network |
| **BAW** | Bulk Acoustic Wave ("TSMR or QMBs" sensors) |
| **CMOS** | Complementary Metal Oxide Semiconductor |
| **COP** | Conducting Organic Polymers (sensor) |
| **CP** | Conducting Polymer |
| **EOF** | Electroosmotic Flow |
| **EPA** | Environmental Protection Agency |
| **ESD** | Electrostatic discharge |
| **FID** | Flame Ionization Detector |
| **FPW** | Flexural Plate Wave |
| **FW** | Feature Weighting |
| **HSSE** | HeadSpace Sorptive Extraction |
| **IC** | Integrated Circuit |
| **IoT** | Internet of Things |
| **LDA** | Linear Discriminant Analysis |
| **MEMS** | MicroElectroMechanical Systems |
| **MISFET** | Metal Insulator-Semiconductor Field-Effect Transistor |
| **MOD** | Machine Olfaction Device |
| **MOS** | Metal-Oxide Semiconductor (sensor) |
| **MOSFET** | Metal Oxide Field Effect Transistors (sensor) |
| **MS** | Mass Spectrometry |
| **MVA** | Multivariate Analysis |
| **PCA** | Principles Component Analysis |
| **PDMS** | Polydimethylsiloxane Silicone elastomer) |
| **PID** | Photolonization Detector |
| **PMMA** | Poly (methyl methacrylate) |
| **PPM** | Parts Per Million |
| **PPy** | Polypyrrole |
| **PTFE** | Poly (tetrtafluoroethylene) |
| **QMB** | Quartz Crystal MicroBalance |
| **SAW** | Surface Acoustic Wave (Sensors) |
| **SBSE** | Stir Bar Sorptive Extraction |
| **SoC** | System on a Chip |
| **SPME** | Solid-Phase MicroExtraction |
| **TSMR** | Thickness Shear Mode Resonators ("BAW" or "QMBs" sensor) |
| **VOCs** | Volatile Organic Compounds |

# Background, materials, and process

In this chapter, the basic outline of a machine olfaction device (MOD) is presented in the form of a discussion and analysis of the various aspects related directly or indirectly to the device. In addition, a brief historical background covering early discussion and development in this field is also presented, while the historical approaches and device mechanisms for detecting and dealing with gases, odor, and aroma from the 20th century and up to recent developments presently taking place have been included as well.

An outline related to problems and solutions relevant to present day's device design is considered as part of various general factors related to the MOD design. Furthermore, sensors and gas systems, plus polymer and manufacturing process, are presented together with other relevant materials connected to the processes and functions of the above device.

## 1.1 Introduction

Within the development of various technologies worldwide, humans are trying, by the application of appropriate technologies, to replicate the function of the five senses they possess. For example, artificial tasting (electronic tongue or eTongue) and artificial smelling (electronic nose or eNose) are some of the modern developments that researchers and technologists at various companies and research institutes are competing for more reliable and accurate output results. MOD presents technology detecting an odor, which can be either for one specific type of substance or for more than one substance, can be achieved only if all the components of the device are working correctly, regardless of the local environment surrounding the device. A good understanding of the chemical structure of a substance can help greatly in achieving this. In a way, the principle of sensing a certain type of odor by MOD can be compared to the outcome of smelling an odor, that is, an airborne polar molecular traveling to our nasal cavity to be sensed by our olfactory system.

The detection threshold by the human nose for different odors can vary widely, depending on the type of compound, that is, some of the substances may have

**Introduction to Machine Olfaction Devices.**
**DOI: https://doi.org/10.1016/B978-0-12-822420-5.00003-9**
© 2022 Elsevier Inc. All rights reserved.

**Table 1.1** Examples of samples odorants properties and structures (Dodd, Bartlett, & Gardner, 1992).

| | Odor type | Threshold in water (ppb) |
|---|---|---|
| | Rose | 290 |
| | Thyme | 86 |
| | Grapefruit | $2 \times 10^{-5}$ |

*Source*: Redrawn and adapted from Dodd, G., Bartlett, N., & Gardner, J., (1992). Odours—The stimulus for an electronic nose. In J. W. Gardner & P. N. Bartlett (Eds.), *Sensors and sensory systems for an electronic nose* (pp. 1–11). Kluwer Academic Publishers, Springer. <https://www.springer.com/gp/book/9780792316930>.

similar chemical formula but generate different smell and vice versa, that is, different chemical formula may generate similar smell—as sensed by humans (Ohloff, 1990) (Table 1.1).

The function of MOD is trying to achieve something similar to the human olfactory system. The questions are "How a machine can measure an odor? How reliable is this kind of measurement? What is needed for repeatable accurate measurement? And what are the benefits from detecting various kinds of odors?" (Box 1.1).

This chapter and the following chapters will provide answers to the above questions and much more. However, in this chapter, a general approach to the MOD system is provided with some additional details, whenever the need arises for further explanations.

## 1.1.1 Brief historical background

In 1914 Alexander Graham Bell was addressing a graduating class of the Friends' School in Washington, DC and during his speech, he made the following statements:

*Did you ever try to measure a smell? Can you tell whether one smell is just twice as strong as another. Can you measure the difference between one kind of smell and another. It is very obvious that we have very many different kinds of smells, all the way from the odor of violets and roses up to asafetida. But until you can measure their likenesses and differences you can have no science of odor. If you are ambitious to found a new science, measure a smell. What is an odor? Is it an emanation of material particles into the air, or is it a form of vibration like sound? If you can decide that, it might be the starting point for a new investigation. If it is an emanation, you might be able to weigh it; and if it is a vibration, you should be able to reflect it from a mirror. Can you reflect a smell or measure its velocity of transmission? If you can do those things you will be well advanced on the road to the discovery of a new science.*

What Bell was describing as a "discovery of a new science" is already an established technology in different parts of the world; however, the historical background related to the MOD actual research did not begin earnestly until the middle of the last century, even though Zwaardemaker and Hogewind did experiment during 1919 related to odor, as they found out that an odorant water solution does have a positive charge compared with pure water (Hogewind & Zwaardemaker, 1919). However, one of the first researches that can be related to MOD is the making of the first gas sensor by Hartman in 1954, followed in 1961 by Mocrief in the form of array of six thermistors. Another attempt such as by Robert Wighton Moncrieffin 1961 was related to a mechanical nose development and in 1964 W.F. Wilkens and J. D. Hartman constructed the first electronic nose. Having said that, the first present approach for electronic nose was not eventually achieved until 1982 by Persaud and Dodd. Further research followed in 1985 via Hitachi Research Laboratory in the form of array for odor quality. In general, sensors and sensor arrays in the form of devices, the name "Electronic Noses" was first associated with them by Gardner and Bartlett (1994).

---

### Box 1.1 Can't we just use dogs?

A dog is very, very sensitive. Special research teams work on training dogs to detect cancers as you would do explosives. What we are trying to do with the electronic nose is create an artificial means of replicating what the dog does. Such machines have the advantage that they do not get tired, will work all day, and you only need to feed them electricity.

Davis (2014)

---

## 1.2 Problems and solutions

The electronic nose or "eNose," presently possesses several problems:

1. Short lifetime of the sensors
2. Sensitivity toward moisture.
3. Narrow selectivity.
4. Relatively expensive.
5. Mostly used as a laboratory tool (many of them are large and nonportable).
6. Results can be difficult to reproduce due to the presence of multiple sensors.

Therefore this body of work is aimed at developing a novel and improved gas sensor with on-board data processing capability for monitoring applications.

One of the systems that is discussed in this book is the use of two sensors, rather than an array of sensors functioning at the same time, as in the case with many eNose devices.

The approach is to develop a simplified field-portable MOD that can have the sensing mechanism and the signal processing components within one unit, for the measurement of volatile species. The device should be able to give real-time analysis that can transmit the results to other devices simultaneously.

This will be achieved by building two sensor devices and exploiting the principles of MOD. Two sensing elements should allow a more reliable and robust device to be constructed as it will also have a greater amenability to size reduction/portability.

The primary objective is to develop a mobile diagnostic system, which can be used in industry/medical environments to monitor certain conditions more accurately than the present systems already available in the market.

Therefore several objectives will be considered, such as:

1. faster stabilizing base line,
2. quicker sample result display,
3. ability to use ambient air,
4. low power consumption, and
5. the ability to deal with different varieties of organic/inorganic samples.

Other areas investigated in this book are mobility, simplicity (how it is easy to use), accuracy (in comparison with other similar devices), cost, and most importantly the design of the device "chamber" and its life cycle (i.e., removable/disposable chamber contents).

## 1.3 MOD design

The idea of producing an MOD to be used for various purposes (commercial and noncommercial) has been around since the first eNose was invented

(Persaud, Dodd, & 1982, 1982). However, in this book, the mobility and longer life sensor(s) of MOD are identified, and a new design has been implemented. To achieve the new design, there are at least three important steps that should be considered first. These are:

1. Some of the present MODs are large, bulky, and too expensive.
2. There are still various problems that affect currently available systems, which is one of the main barriers delaying its use more often within our daily needed tasks, such as within the commercial and industrial setting, as well as within the domestic market.
3. How reliable and accurate is the system, what kind of standardization will be required?

Therefore the principle in this book is to design a small mobile MOD, while at the same time develop its functions for better and accurate performance and longer life cycle for the device and related functionality. The specialized parts of the above proposed MOD design are discussed in detail in this book.

## 1.4 Why this book?

Present MODs need more development and adjustment to bring them to the level where they can be reliable and easy to use and not restricted to the realms of laboratories and commercial businesses. This can be achieved by implementing the above five points mentioned under "Objectives." Therefore producing a small mobile device is the next appropriate step due to the growing demand worldwide for specific applications. The same thing can be said about the final application of the device itself, which can be within any specialized area, as in the case of enclosed environmental issues as well as open environment.

An outline below will provide some understanding concerning the work provided in this book.

## 1.5 Background and current state of technology

Large number of books and articles concerning MODs have been written and published, mostly connected to the food and beverage industry or certain medical connection. However, there are other areas of interest as well, where applications of the MOD can benefit society as a whole:

*Industrial*: Online process control and production, QA/QC activities within a large range of the industrial sectors.

*Environment*: Monitoring environmental pollution in general.

*Fraud*: Identifying unauthorized copies and imitation for various goods/products (e.g., wines, spirits to perfumes and cosmetics).

*Law*: Detection of vapors from explosives and drugs.

*Health and safety*: Detection of bacteriological and chemical contamination.

The need for a better and more efficient method to detect volatile organic compounds (VOCs) accelerated the research and development of the MOD devices. For example, there are already several MOD devices developed to monitor the degradation of oil. However, monitoring online in any area by using a mobile device is a more efficient way in obtaining an instant result on a 24-hour basis, that is, whenever and wherever the need arises for this kind of monitoring the device is available to do the work. Developing efficient, reliable, and low-cost mobile MOD device is part of the present research taking place across the world.

---

## Box 1.2 Characterization of odors

The characterization of odors with an electronic nose is divided into preprocessing and pattern recognition. Preprocessing typically involves the calculation of relative or fractional changes in each parameter, to compensate for noise and drift, and some form of normalization. A Bayesian or neural network classifier then exploits the distributed representation of the normalized sensor responses as high-dimensional vectors.

Denham (2006)

---

As mentioned earlier, there are presently various types of devices to detect gases and vapor/odors (Boxes 1.2 and 1.3) such as gas detectors and devices with single or multiple sensors. They all work under the same principles of measuring a change of current resistance (Stetter & Penrose, 2001).

Any MOD device is usually made up of three (or more) important parts, a single or an array of chemical sensors, delivery system, and a pattern-recognition system.

A single, or arrays of chemical sensors supply a set of measurements to a pattern-recognition system. The pattern-recognition system compares the pattern of the measurements obtained to stored patterns of similar materials already known to the device itself. All the films on a set of electrodes (sensors) start out at a measured resistance, called the sensor's baseline resistance. If there is no change in the composition of the air, the films stay at the baseline resistance and the percentage change is zero. If there is a change, then an identification of the sample can be made. When the above two systems are working correctly, we have what is called MOD device (which can be a single or multiple sensor systems).

As further global research is successfully completed in this field, many of the common problems have been already solved either completely or provided with partial solutions. For example, problems related to signal drift, humidity, temperature, and air quality. Some of these problems have been addressed by designing higher quality sensors [e.g., electro-spray for quartz microbalance (QMB) sensor production] and better performing chambers (see Chapter 5: Biosensors and MOD Design, for details). In addition to the above, new sampling techniques, for example, solid phase microextraction/stir bar sorptive extraction helped to obtain better results in the analysis of semivolatile (odoriferous) compounds as part of the MOD analysis already established (Ampuero & Bosset, 2003).

---

## Box 1.3 Aromas

Aromas are simple to complex mixtures of volatile compounds present in the air at concentration that may be detected by animals through the sense of olfaction. Aromas sometimes have been referred to as "smells" or "odors" when a particular connotation referring to the pleasantness or unpleasantness of an aroma is being expressed. In some cases, the aroma is composed of a single chemical compound, while in others only a few compounds may be present of which only one may be the dominant or principal component.

Wilson and Baietto (2009)

---

## 1.6 Gas sensing systems

### 1.6.1 Sensors

Since this book is dealing with a mobile version of MOD system, it will be good practice to mention some examples of miniaturized/mobile MOD (handheld instruments). These handheld devices can perform basic gas sampling procedures; therefore they are limited in their functions. They depend on microprocessors (embedded software system) for system control and data processing. Fig. 1.1 shows various types of gas sensing systems and their relationship to each other.

There are several different types of sensors that can be used as an essential component of different designs of MOD systems, the following are few examples: (1) electrochemical sensors, (2) metal oxide semiconductors (MOS), (3) Schottky diode—based sensors, (4) calorimetric sensors, (5) quartz crystal microbalance (QCM), and (6) optical sensors.

MOD sensors fall into five categories:

> conductivity sensors
> piezoelectric sensors
> metal oxide field effect transistors
> optical sensors
> spectrometry-based sensing methods

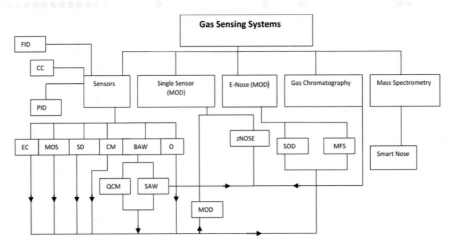

**Figure 1.1** Various gas sensing systems and how they are related to each other (sensors' relationship). CC, Catalytic combustion, CM, calorimetric, EC, electrochemical, FID, flame ionization detector, MFS, mass-flow systems, MOS, metal oxide semiconductor, O, optical (there are also smart sensors not mentioned above simply because they are like any of the above sensors with a processor added to the sensor itself), PID, photoionization detector, QCM, quartz crystal microbalance, SD, Schottky diode based, SOD, static odor delivery.

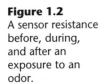

**Figure 1.2**
A sensor resistance before, during, and after an exposure to an odor.

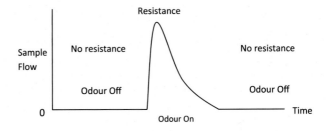

**Table 1.2** Sensor devices showing the physical changes in the sensor active film.

| Physical changes | Type of sensor |
| --- | --- |
| Conductivity | Conductivity sensors |
| Mass | Piezoelectric sensors |

Conductivity sensors metal oxide and polymer exhibit a change in resistance when exposed to VOCs (Nagle, Schiffman, & Gutierrez-Osuna, 1998) (Fig. 1.2 and Table 1.2).

MOS, conducting polymer (CP), and QCM devices will be examined in detail, as they are well researched, documented, and established as an important part of various types of MOD devices. The area of application for the proposed device—which will be trained to analyze—will greatly influence the choice of sensor. Regarding the work in this book, either one application at a time can be implemented or various applications can be applied at the same time.

According to Hurst (1999), the response of the sensor is a two-part process:

1. The vapor pressure of the analyte usually dictates how many molecules are present in the gas phase and consequently how many of them will be at the sensor(s).
2. When the gas phase molecules are at the sensor(s), these molecules need to be able to react with the sensor(s) to produce a response.

Sensors types used in any MOD can be mass transducers, for example, QMB or chemo-resistors, which are based on metal oxide or CPs. In some cases, arrays may contain the above two types of sensors. The following sensors are discussed only briefly, as they may not form an important part in relation to the work of this book: catalytic combustion (CC), photoionization detector (PID), flame ionization detector (FID). However, because of their functions in detecting the make-up of a sample, similar in principle to any MOD device, it is a good practice to include them here.

## 1.6.2 Catalytic combustion

Usually, this device is employed for safety applications because of its reliability and simplicity of use. Its response is based on thermal changes, such as those coming out from the combustion of the gas that the device is sensing (Boxes 1.4 and 1.5).

### Box 1.4 Catalytic combustion

CC is receiving a great deal of attention because it has the ability to reduce pollutants to levels far below that which can be achieved by the systems described previously. Fuel and air are mixed thoroughly before entering the catalyst, which promotes chemical reactions and therefore releases the heat of combustion. CC can take place at equivalence ratios that are well below the lean extinction limit encountered in conventional combustion systems. At such low temperatures, the NOX levels are reduced dramatically. An intermediate zone is provided to convert any CO and UHC into products such as $CO_2$ and $1-420$ and is followed by a dilution zone to prepare the combustor exhaust gases for entry into the turbine section.

At start-up and idle, the compressor delivery temperature may be too low for the catalyst to be effective and a separate combustion chamber may be needed for start-up and during idle operating conditions. CC is still under development. The

*(Continued)*

---

## Box 1.4 (Continued)

significant problems to overcome are satisfactory catalyst life and reliability in the harsh and varied operating conditions that prevail in a gas turbine combustor.

It must be pointed out that the turbine entry temperatures (TET) have increased progressively and gas turbines today operate at firing temperatures of 1800K. At such high temperatures, the potential for emission reduction using CC is limited, as stable combustion is possible with other forms of the low emission combustion systems discussed earlier. Since these combustion systems are quite well developed, CC is most likely to find application in small units where the TET is below the weak extinction limit.

However, if the control system for CC is significantly simpler than the DLE combustion system (particularly if overboard bleeds are dispensed with), then there may be a strong case for widespread use of the CC in gas turbines, provided the cost of such combustion systems are competitive. It is also worth pointing out that the TET is unlikely to exceed 1800K because, at higher TET, NOX emissions increase significantly.

Razak (2007)

---

*Advantages*

1. Uniform responses to a variety of flammable gases/vapors
2. Simple to use and easy to calibrate
3. Less sensitive to humidity and changes in air
4. Able to detect most hydrocarbon gases

*Disadvantages*

1. Lower detectable limit
2. Sensor poisoning
3. Requires oxygen to operate

## 1.6.3 Photoionization detector

For the aim of removing an electron, usually this depends on the energy required for this purpose; therefore a PID can detect a compound via ionization. This will happen when the PID's lamp energy is higher than the compound's ionization potential, then the PID will be able to detect it. PID is a portable vapor and gas detector, mainly for the detection of organic compounds. As mentioned above, photoionization happens when either an atom or a molecule is exposed to light that causes an electron to leave, that is, results in the creation of a positive ion.

*Advantages*

1. Low cost.
2. Ease of operation.
3. Results obtained within very short time.
4. Ability to detect low level of VOCs—around 100 parts per billion (ppb).

*Disadvantages*

1. Small monitoring ranges.
2. Limited detectable compounds.
3. Susceptibility to interference from water vapor.
4. PID lamp can gradually lose power even if not in use.
5. Many PID lamps have a life expectancy of less than 1 year.
6. Not suitable for the detection of semivolatile compounds.
7. Requires recalibration frequently.

## 1.6.4 Flame ionization detector

The measuring of compounds is done by utilizing a flame produced by the combustion of hydrogen and air. Ions are produced when hydrocarbons in the sample are brought into the detection zone.

In comparison with CC and PID, the FID has more advantages than the previous two devices, such as:

1. Relatively low detectable limit (200 ppb or less).
2. Wider detecting range.
3. Sensitive to all hydrocarbon vapors.
4. Stable and repeatable.
5. Fast recovery and response times.
6. Unaffected by water vapor or ambient level of CO and $CO_2$.
7. The FID responds better with molecules that contain only carbon and hydrogen.

---

### Box 1.5 Catalytic combustion sensors

CC sensors are classified according to the method employed to measure combustible gas, and they have come to be widely used to prevent methane explosions in coal mines from about 1959. Since then, they have been improved considerably and are now more widely used throughout various industrial sectors as reliable sensors.

- Accuracy and reproducibility are excellent.
- Low power consumption allows devices to be more compact.

*(Continued)*

---

> **Box 1.5 (Continued)**
> - Hardly affected by ambient temperature and humidity.
> - The output curve (to just short of the explosion threshold) is almost a straight line.
>
> GASTEC (undated)

### 1.6.5 Metal oxide semiconductors

These sensors were originally produced in Japan in the 1960s and were used in gas alarm devices, for example, $SnO_2$, $ZnO$, $WO_3$, and $TiO_2$.

MOSs have been used more extensively in eNose instruments and are widely available, commercially. These sensors are made of a ceramic heated by a heating wire and coated by a semiconducting film. They can sense gases by monitoring changes in the conductance during the interaction of a chemically sensitive material with molecules that need to be detected in the gas phase. Out of many MOS, the material that has been experimented with most is "tin dioxide—$SnO_2$"; this is because of its stability and sensitivity at lower temperatures (Sberveglieri, 1999). Different types of MOS may include oxides of tin, zinc, titanium, tungsten, and iridium, doped with a noble metal catalyst such as platinum or palladium.

In general, there are issues that MOS may exhibit, such as:

1. Direct interference in the form of environmental factors, for example, temperature, humidity.
2. Temporal shift of the sensor(s), that is, drift.
3. Under similar conditions, two (or more) sensors may provide different outputs, that is, signal shift and baseline differences in measurement data.
4. Susceptibility to interference from other odors from surrounding environment that is not part of the test.

MOSs are subdivided into two types: MOS with thick film and MOS with thin film.

*Limitation of the thick film MOS:* Less sensitive and poor selectivity, it needs a longer time to stabilize, higher power consumption. This type of MOS is easier to produce and therefore costs less to purchase.

*Limitation of the thin film MOS:* Unstable, difficult to produce and therefore more expensive to purchase. On the other hand, they have much higher sensitivity and much lower power consumption than the thick film of the MOS device.

### 1.6.5.1 Manufacturing process

Polycrystalline is the most common porous material used for thick film sensors. It is usually prepared in a process called "sol−gel," explained briefly as follows.

Stannic chloride ($SnCl_4$) in aqueous solution, then adding ammonia to precipitate tin tetra hydroxide and this precipitate tin tetra hydroxide is dried and later calcinated at 500°C−1000°C. The resulting tin dioxide powder is grounded and mixed with dopands (metal chlorides).

To remove the chlorine as well as getting the pure metal, the powder is heated.

For the purpose of screen printing, paste is made up from the powder.

Finally, with a layer of few hundred microns, the paste is left to cool (e.g., on a tube alumina or plain substrate) (Nose Office, 2003).

### 1.6.5.2 Sensing mechanism

The change of conductance in the MOS is the basic principle of the operation in the sensor itself. That means the change in conductance takes place when an interaction with a gas happens. The conductance will vary depending on the amount of concentration of the gas itself.

#### 1.6.5.2.1 Metal oxide sensors types
1. n-type [zinc oxide, tin dioxide, titanium dioxide iron (III) oxide]
2. p-type (nickel oxide, cobalt oxide)

The n-type usually responds to reducing gases, while p-type responds to "oxidizing" gases.

#### 1.6.5.2.2 Operation (n-type)
As the current is applied between the two electrodes, via the metal oxide, oxygen in the air will start to react with the surface and accumulate on the surface of the sensor, consequently *trapping free electrons on the surface from the conduction band* (Arshak, Moore, Lyons, Harris, & Clifford, 2004). In this way, the electrical conductance decreases as resistance in these areas increase due to lack of carriers (i.e., increase resistance to current), as there will be "potential barriers" between the grains (particles) themselves. When the sensor is exposed to reducing gases (e.g., CO) the resistance drops as the gas usually reacts with the oxygen and therefore an electron is released. Consequently, the release of the electron increases the conductivity as it will reduce "the potential barriers" and let the electrons to start to flow.

#### 1.6.5.2.3 Operation (p-type)
Oxidizing gases (e.g., $O_2$, $NO_2$) usually remove electrons from the surface of the sensor, and as a result of this, charge carriers will be produced (Figs. 1.3 and 1.4).

**Figure 1.3**
Sensors made of a
ceramic, heated
and then coated
by a
semiconducting
film. *Source:*
*Redrawn from*
*Sberveglieri, D.*
*(1999). Metal-oxide*
*semiconductors. In:*
ASTEQ technologies
for sensors.

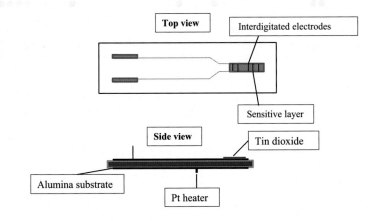

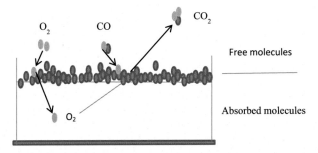

**Figure 1.4** As the current is applied between the two electrodes, via "the metal oxide," oxygen starts to accumulate on the surface of the material, consequently "trapping electrons from the conduction band." In this way, the electrical conductance decreases. *Source: Redrawn from Sberveglieri, D. (1999). Metal-oxide semiconductors. In: ASTEQ technologies for sensors.*

### 1.6.5.3 Limitation of MOS sensors

1. Poor selectivity—When a thick film MOS device is used, the poor selectivity can be reduced by adding a deposition of a suitable catalyst layer of noble metals. Pd, Pt, Au, and Ag are sometimes added for this purpose.
2. MOS need high temperatures (around 300°C) to operate efficiently; this means higher power consumption is needed.
3. Sensitive to humidity and to compounds, for example, ethanol and $CO_2$.

### 1.6.5.4 Advantages

1. Widely available in the market in a variety of types and sensitivities.
2. Very sensitive to a number of organic vapors (e.g., oil).
3. Fast response, usually in less than 10 seconds.

### 1.6.6 Conducting polymers

A polymer is basically a substance made up of many repeating chemical units (or molecules). CPs, as the name indicates, are conjugated polymers, namely organic compounds that have an extended p-orbital system (Fig. 1.5), through

**Figure 1.5**
Schematic illustration of a cross-section of three-dimensional region of space of p-orbital system.

which electrons can move from one end of the polymer to the other. The most common are polyaniline and polypyrrole (MacDiarmid, 1997).

A CP film is usually used as a sensor to detect vapor/odor under the same principles as in MOS. Polymers can be used for many devices combining unique optical, electrical, and mechanical properties. CPs therefore are used for their optical effects and underlying physical processes.

Polymers can be divided into the following categories: organic polymers, copolymers, conjugated polymers, for example, poly(*para*-phenylene), polyaniline, and poly(*para*-phenylenevinylene), characterized by high flexibility (Box 1.6). Most CPs can be made to transfer electrons to other materials such as Buckminsterfullerene *(Carbon 60 C60 Buckyball).*

Both inorganic and organic materials can be used to create light-emitting diodes, such as InGaN materials or cadmium selenide nanocrystals, where the physical process involves quantum wells.

CPs are made by "electro polymerization" of complex organic dyes—specifically derivatives of the substances polypyrole, polyaniline, and polythiophenes. Depending on the exact chemical structure of the polymer (Fig. 1.6), each one can be given a different conductive behavior. In this way, a list (or database library) can be built for different types of CP (i.e., sensors) with each one testing a different type of molecular (MacDiarmid, 1997).

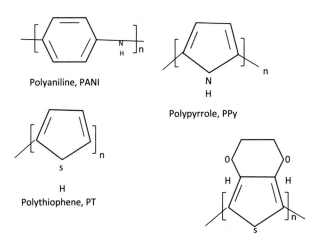

Polyaniline, PANI

Polypyrrole, PPy

Polythiophene, PT

Poly(3,4-ethylenedioxythiophene), PEDOT

**Figure 1.6** Examples of chemical structure of representative conducting polymer (Nguyen & Yoon, 2016). *Source: Redrawn from the source Nguyen D., & Yoon, H. (2016).* Recent advances in nanostructured conducting polymers: From synthesis to practical applications. *MDPI. Open Access. <https://www.mdpi.com/2073-4360/8/4/118/htm>.*

## Box 1.6 Conducting polymers

Most organic polymers are insulators by nature. However, a few intrinsically CPs exist that have alternating single and double bonds (*conjugated bonds*) along the polymer backbone or that are composed of aromatic rings such as phenylene, naphthalene, anthracene, pyrrole, and thiophene, which are connected to one another through carbon − carbon single bonds.

The first polymer with significant conductivity synthesized was polyacetylene (polyethyne). Its electrical conductivity was discovered by Hideki Shirakawa, Alan Heeger, and Alan MacDiarmid who received the Nobel Prize in Chemistry in 2000 for this discovery. They synthesized this polymer for the first time in the year 1974 when they prepared polyacetylene as a silvery film from acetylene, using a Ziegler − Natta catalyst. Despite its metallic appearance, the first attempt did not yield a very conductive polymer. However, 3 years later, they discovered that oxidation with halogen vapor produces a much more conductive polyacetylene film. Its conductivity was significantly higher than any other previously known conductive polymer. This discovery started the development of many other conductive organic polymers.

CROW Polymer Science (undated)

### 1.6.6.1 Polymer preparation

CP sensors are made by chemical or electrochemical polarization from monomers, such as aniline or pyrrole. The addition of dopants (they can be any kind of conductive materials) increases the polymer conductivity as they create an accumulation of positive or negative particles, for example, self-doped polyaniline.

### 1.6.6.2 Sensing mechanism

When the analyte interacts with the sensing surface (when molecule interacts with another molecule) the resulting output is a detectable signal. This is the basic principle of how a chemical sensor works. All polymers, in general, have a similar detecting mechanism. Chemical sensors based on conjugated polymers detect a variety of analytes as they can detect analyte at low concentrations. This is because they contain a "Chromophore," which is requirement for energy for the excitation of an electron is very low (Wackerly, 2004).

When contact is made with analyte molecules the conductivity of the "polymer" changes, and a current will be created within the sensor in proportion to the concentration of the new substance or analyte. The next step is that the generated current will usually be detected by the signal processing circuits in

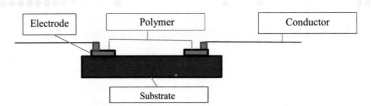

**Figure 1.7** Sensors obtained by electropolymerization of a film of polymer between gold-plated electrodes—more details in the appendix (Vaefolomeev, 1999). *Source: Redrawn from Vaefolomeev, S. (1999). Conducting polymer sensors. In: ASTEQ technologies for sensors.*

the MOD (Fig. 1.7). A pattern will then be generated indicating the type of element/molecules in the sample.

### 1.6.6.3 Polymer advantages

1. They work within a wide range of operating conditions (e.g., variations in average room temperature, pressure, and humidity).
2. Functional groups that interact with different classes of odorant molecules can be built into active material, providing high sensitivities.
3. Organic materials tend to be easier to use than oxides as they can be used close to room environmental conditions than, for example, MMOS. Also, they are more easily modified to react with specific gaseous species than inorganic materials.

### 1.6.6.4 Polymer disadvantages

1. Even though polymer can be made highly sensitive (see point 2 above), organic materials in general are usually poor conductors, hence measurement conductivity can be difficult.
2. Can be thermally unstable.
3. Can be sensitive to water.

## 1.6.7 Quartz crystal microbalance

The QCM is an extremely sensitive mass sensor, capable of measuring mass changes in the nanogram range. QCMs are piezoelectric devices fabricated of a thin plate of quartz with electrodes affixed to each side of the plate. A QCM-D (QCM with dissipation monitoring) consists of a thin quartz disk sandwiched between a pair of electrodes.

Due to the piezoelectric properties of quartz, it is possible to excite the crystal to oscillation by applying an AC voltage across its electrodes. Changes to this oscillation are directly proportional to mass changes on the crystal. Various sorbent coatings can be used on the crystal surface to determine the selectivity of the sensor. Several types work under similar basic principles, such as "Bulk Acoustic Wave (BAW)" and "Surface Acoustic Wave (SAW)" sensors. Both sensors need AC voltage for configurations; however, BAW sensors use the electric field to excite the quartz crystal to oscillate, and SAW sensors use wave propagation on the sensor surface.

### 1.6.7.1 Manufacturing process

After being cut along certain crystallographic axis, the thin plates of the single piezoelectric crystal quartz will be covered with thin gold electrodes on both sides. Then the two sides are coated with polymer films. The coating techniques could be in any of the following methods:

1. spray coating
2. growth of Langmuir − Blodgett films
3. self-assembled monolayers

The coating will provide the conductivity and changing of mass.

### 1.6.7.2 Sensing mechanism

The QCM is basically a thin quartz wafer with electrode pads on each side. The QCM oscillates in a mechanical way as and when connected to an amplifier. At the same time the amplifier oscillates electronically, with a certain frequency.

On the surface of QCM there is a deposit of a sensitive chemical coating. Exposing the coating to analyte vapor, the molecules of the analyte enter into the coating. The result will be an increase in mass, which causes a slowing in the oscillation frequency. QCM is very sensitive to any minute changes in its mass, for this reason the QCM can measure changes in its frequency to 1 part in $10^8$; however, normal operating frequencies are in the range from 10 MHz to 30 MHz (Nose Office, 2003).

## 1.6.8 Surface acoustic wave sensors

As in the QMB (i.e., QMC) this sensor is based on the same principle, that is, when mass changes, frequency changes. The device uses SAWs, with a frequency of about 600 MHz.

### 1.6.8.1 Manufacturing process

Two interdigital transducers (IDT) are usually made up from thin metal electrodes and fitted on a polished piezoelectric substrate, located in the center and enclosed by resonators (Nose Office, 2003). The wavelength is determined by the spacing of the IDT fingers. One of the IDT surfaces will expand and contract when an alternating current is applied. The movement of the surface generates a wave, some scientists/researchers call it Rayleigh waves, which will pass through the substrate. On the other hand, the frequency counter of the IDT receiver will record the frequency.

To minimize noise and temperature, as well as to lower the frequency to be measured, a dual SAW setup is constructed; therefore the reference signal from SAW (uncoated) will be mixed with the sensor signal.

### 1.6.8.2 Sensing mechanism

The physical properties of the surface can affect the wavelength/frequency of the surface wave itself. A thin layer of polymer coats the substrate, which is located between the two IDTs. The absorption of gas changes the mass and consequently the properties of the sensitive layer. The surface wave is not just affected by the change of mass; it is affected by four other factors, such as temperature, pressure, dielectric constant, and viscosity.

## 1.6.9 Smart sensors

Smart sensors are simply sensors with microprocessors attached to them. This may include signal processing and feedback are integrated with the sensor itself, as well amplification of the sensor output to digitization and, therefore it can stop or reduce the noise in the transmission link.

When it comes to a system design, a smart sensor can be:

easier
cheaper
more reliable and more scalable
higher performance
faster in designing a sensor system

Obviously, these benefits are all obtained when microprocessors or computing resources are embedded on the sensor. Therefore the processing of data is performed on the spot, that is, within each individual sensor, instead of using a central system controller. In addition to this, ordinary sensors output raw data; only useful data are produced by a smart sensor. Many of the smart sensors can be easily programmed and/or reprogrammed, thus saving time and expenses. Smart sensor's disadvantages are mainly related to higher costs and to the complexity of the sensor system.

The feasibility of using such kind of sensors for new MOD depends on how small the device will be and on the final application(s), as well as the final cost of the device itself. Having said that, the design for the new MOD proposed in this book, there will be other types of smart sensors within the device. These smart sensors are mainly for the purpose of monitoring (if not these monitoring functions are already built within the main sensors of the device), for example, internal and/or external related to humidity, change of temperature, atmospheric pressure etc., that is, within the device and outside it and for the purpose of informing the user in advance; and whenever it is possible, to adjust the outcome according to the new changes in the environment that may have impact directly or indirectly on the function of the MOD in general, and the accuracy of the final results, in particular. The device can provide a report and/or possible percentage of error, if the device is used directly under the above new conditions (Box 1.7).

---

### Box 1.7 Smart sensors

The smart sensor possesses several functional layers: signal detection from discrete sensing elements, signal processing, data validation and interpretation, and signal transmission and display. Multiple sensors can be included in a single smart sensor system whose operating properties, such as bias voltage or temperature, can be set by the microprocessor. The sensor elements interface to signal smart sensor systems by Gary W. Hunter, Joseph R. Stetter, Peter J. Hesketh, Chung-Chiun Liu control and conditioning stages that will provide both excitation and signal data logging and conditioning. The data acquisition layer will convert the signal from analog to digital and acquire additional parameters of interest to provide compensation when needed for thermal drift, long-term drift, etc. The embedded intelligence will continuously monitor the discrete sensor elements, validate the engineering data being provided, and periodically verify sensor calibration and health. The processed data become information and can then be transmitted to external users. The user can choose the complexity of the data transmitted: from a single reading to a complete download of the sensor system's parameters.

(Hunter, Stetter, Hesketh, & Liu, 2010)

---

## 1.7 Dynamic olfactometry

Dynamic olfactometry is a method for providing the level of concentration of an odor in a sample via the sensation of correctly selected panel of people (referred to "sensorial techniques") where the odor of the above sample is directly exposed to them and measured in European odor units per cubic meter ($ouE/m^3$), as has been standardized by the European Standard EN13725:2003 (Bax, Sironi, & Capelli, 2020).

*European Standard specifies a method for the objective determination of the odour concentration of a gaseous sample using dynamic olfactometry with human assessors and the emission rate of odours emanating from point sources, area sources with outward flow and area sources without outward flow. The primary application is to provide a common basis for evaluation of odour emissions in the member states of the European Union.*

According to European Standard (2003), the field of application of this European Standard includes the following: (1) The measurement of the mass concentration at the detection threshold of pure odorous substances in $g/m^3$. (2) The measurement of the odor concentration of mixtures of odorants in $ouE/m^3$. (3) The measurement of the emission rate of odorous emissions from point sources and surface sources (with and without an outward flow), including predilution during sampling. (4) The sampling of odorants from emissions of high humidity and temperature (up to 200°C). (5) The determination of effectiveness of end-of-pipe devices used to reduce odor emissions.

However, the European Standard excludes the following: (1) The measurement of odors potentially released by particles of odorous solids or droplets of odorous fluids suspended in emissions. (2) The measuring strategy to be applied in case of variable emission rates. (3) The measurement of the relationship between odor stimulus and assessor response above detection threshold. (4) Direct measurement of hedonic tone [or (un)pleasantness] or direct assessment of potential annoyance. (5) Field panel methods. (6) Measurement of recognition thresholds.

Also, DO is regulated by the UNI EN 13725: 2004 Technical Standard (Table 1.3).

**Table 1.3** Example of odor regulations criterion in certain countries (Deshmukh, Bandyopadhyay, Bhattacharyya, Pandey, & Jana, 2015).

| Country | Odor criterion | | Remarks |
|---|---|---|---|
| Denmark | 0.5–10 ou m$^3$ | | 99 percentile, with an averaging time of 1 min |
| Ireland | 1.5–6 ou m$^3$ | | Averaging time 1 h |
| The Netherlands | 0.5–3.5 ou$_E$ m$^3$ | | Averaging time 1 h |
| South Korea | 15–20 OC | | |
| New Zealand | 1–5 ou m$^3$ | | 99 percentile, with an averaging time of 1 h |
| Australia | 2–10 ou m$^3$ | | Averaging time varies from 3 min to 1 h |
| Taiwan | 50 ou m$^1$ | | |
| United States | 4–50 DT$^1$ H$_2$S | 0.5 ppb | Average time varies with reference to source and surrounding residential area 30 min averaging window |
| Germany | 0.1 ou m$^3$ | | 10% of the hours for residential areas, 15% for industrial areas, values at the emissions must be kept at 200 ou m$^3$ |

*Source*: Edited from Deshmukh, S., Bandyopadhyay, R., Bhattacharyya, N., Pandey, R., & Jana, A. (2015). *Application of electronic nose for industrial odors and gaseous emissions measurement and monitoring — An overview* (Vol. 144, pp. 329–340). Elsevier. <https://www.sciencedirect.com/science/article/pii/S0039914015300904>.

# 1.8 Sensor drift

Sensor drift can be a problem depending on the type of sensor films, sensor architectures, and the length of time the senor being used (i.e., age). Why is drift a problem? If it is not possible to repeat the same result on the same sample under the same conditions, then drift can be one of the causes.

Sensor drift can be caused as a result of environmental disturbance (temperature, humidity, and variation in ambient pressure—referred to as application variables), sensor (surface) aging, or sensor poison, or in certain cases it could be that the chemical reaction inside the sensors affected when exposed to some odorant.

The problem can usually be corrected either statistically, through calibration (regular calibration to the sensor), improvement on the method and materials in manufacturing the sensors, or by simply replacing the old sensor with a new one (Box 1.8).

Questions may arise on the methods of preventing or more likely minimizing sensor drift! The first one is how often calibration should be made to the device sensor(s)? Obviously, whenever the result indicates a possible error, either the error is generated/indicated via the device itself or via comparison with established data or comparison with other devices. In general, regular calibration is required on many present devices; however, recent designed MOD devices may not require this kind of regular calibration. Depending on how often the device is used, where stored, and the type of sensor(s) as well as related other hardware components the device constructed from, plus the device application (s) programs, all play role in sensor drift. The design of the MOD in this book is aiming for a calibration once or maximum twice a year, if used within an average normal condition, as specified within the manual instructions of the device.

The installation of the sensor when replacing a faulty one can also be the source of errors if not installed correctly. For this reason, this should be checked before deciding on calibration of the device. However, it is possible to separate drift from the analytical signal and modeled via a method called "drift compensation and modeling method" by obtaining a model that can be referred to to correct the senor drift outcome via new samples. The above approach, which is named "component correction or CC," is that when drift takes place, it is noticed that there is a correlation in sensor's response when this happens, as the drift of the sensors point to a certain direction, similar to the tested various samples, as well as the reference gas, that is, by knowing when the drift occurs and providing the direction of the drift, such as in the form of a model (in the reference samples) then subtracting it from the latest data obtained (i.e., drift data) (Artursson et al., 2000).

## 1.8.1 Calibration transfer

As the conditions differ when MOD is used in one particular field (e.g., A) compared with the conditions in a new environment (e.g., B) (this may include open and enclosed environment), where the training provided in (A) differ from the environment of (B) then difficulties may arise in obtaining accurate results. Sensor drift can be part of the above issues as well other issues such as different baseline conditions between (A) and (B), samples from (A) may have different device training than those for samples from (B), calibration standard may require to be transferred to a different device (D) and when replacing a faulty sensor. Therefore calibration transfer may help to solve some, if not all, of the above issues. This can be done via multivariate implementation (Nakamoto, 2016).

---

## Box 1.8 Drift phenomenon in metal oxide semiconductor gas sensors

Among the different technologies used to fabricate chemical sensors, which in general lack enough information about the physical causes of the sensor drift, the physical meaning of the drift phenomenon in conductometric metal oxide sensors (MOX) has been deeply investigated in the past. In fact, signal drift is known to be a severe problem for these devices, which have widespread commercial diffusion. A typical example is the sintered tin dioxide Taguchi Gas Sensor (TGS), an n-type semiconductor solid state device marketed by Figaro Engineering Inc.2 since 1968 and widely applied for detection of oxidizing and reducing gases. The sensing properties of TGS are based on the electronic (n-type) conductivity of tin dioxide ($SnO2$). The device consists of small $SnO_2$ grains, which are in contact with each other. The sensing effect is due to an electronic depletion layer at the surface of the grains. The depletion layer is generated when oxygen is adsorbed, thus trapping electrons from the oxide. This induces an increased resistance at the grain surfaces. When the current passes from one grain to another it has to cross these depletion layers, which thus determine the sensor's resistance. The sensing effect, that is, the response to reducing or oxidizing gases, is therefore determined by the adsorption of these compounds and the subsequent trapping (or donation) of electrons by the adsorbed species. This modifies the space charge potential thus changing the sensor's conductivity. Similar working models also apply to thick- and thin-film semiconductor MOX gas sensors, including for instance $SnO_2 - RGTO3$ gas sensors.

Carlo and Falasconi (2012).

---

## References

Ampuero, S., & Bosset, J. O. (2003). The electronic nose applied to dairy products: A review. *Sensors and Actuators B, 94*, 1−12.

Arshak, K., Moore, E., Lyons, G. M., Harris, J., & Clifford, S. (2004). A review of gas sensors employed in electronic nose applications. *Sensor Review, 24*(2), 181−198.

Artursson, T., Eklöv, T., Lundström, I., Mårtensson, P., Sjöström, M., & Holmberg, M. (2000). Drift correction for gas sensors using multivariate methods. *Journal of Chemometrics, 14*, 711−723.

Available from https://doi.org/10.1002/1099-128X(200009/12)14:5/6 < 711::AID-CEM607 > 3.0. CO;2-4. <https://onlinelibrary.wiley.com/doi/10.1002/1099-128X(200009/12)14:5/6%3C711:: AID-CEM607%3E3.0.CO;2-4 >.

Bax, C., Sironi, S., & Capelli, L. (2020). How can odors be measured? An overview of methods and their applications. *Atmosphere, 11*, 92. Available from https://doi.org/10.3390/atmos11010092.

Carlo, D., & Falasconi, M. (2012). Drift correction methods for gas chemical sensors in artificial olfaction systems: Techniques and challenges. In W. Wang (Ed.), *Advances in chemical sensors*. InTech. ISBN: 978-953-307-792-5 <https://core.ac.uk/download/pdf/11421334.pdf >.

CROW Polymer Science. (undated). *Polymer properties database—Conducting polymers.*

Davis, N. (2014, April 2). Interview: Electronic nose explained: In future we will be sniffing out disease. *The Guardian Newspaper*, UK Edition. <https://www.theguardian.com/science/2014/apr/02/electronic-noses-explainer-sniffing-disease>.

Denham, M. (2006). Sensory processing. In L. Tarassenko, & M. Denham (Eds.), *Cognitive systems—Information processing meets brain science*. ScienceDirect—Elsevier.

Deshmukh, S., Bandyopadhyay, R., Bhattacharyya, N., Pandey, R., & Jana, A. (2015). *Application of electronic nose for industrial odors and gaseous emissions measurement and monitoring — An overview* (Vol. 144, pp. 329–340). Elsevier. <https://www.sciencedirect.com/science/article/pii/S0039914015300904>.

Dodd, G., Bartlett, N., & Gardner, J. (1992). Odours—The stimulus for an electronic nose. In J. W. Gardner, & P. N. Bartlett (Eds.), *Sensors and sensory systems for an electronic nose* (pp. 1–11). Kluwer Academic Publishers, Springer. <https://www.springer.com/gp/book/9780792316930>.

European Standard. *EN 13725:2003/AC*. (2003). <https://infostore.saiglobal.com/preview/is/en/2003/i.s.en13725-2003%2Bac-2006.pdf?sku=658463>.

Gardner, J., & Bartlett, P. (1994). A brief history of electronic noses. *Sensors and Actuators B: Chemical, 18*(1–3), 210–211. Available from https://www.sciencedirect.com/science/article/abs/pii/0925400594870853.

GASTEC. *Gas sensors—Catalytic combustion sensors*. (undated). <https://www.gastec.co.jp/en/product/detail/id = 2205>.

Hogewind, F., & Zwaardemaker, H. (1919). On spray-electricity and waterfall-electricity. In *KNAW proceedings* (Vol. 22 I, pp. 429–437), Amsterdam.

Hunter, G. W., Stetter, J. R., Hesketh, P. J., & Liu, C.-C. (2010). *Smart sensor systems*. The Electrochemical Society Interface Winter. <https://www.electrochem.org/dl/interface/wtr/wtr10/wtr10_p029-034.pdf>.

Hurst, W. J. (1999). *Electronic noses & sensory array based systems*. Technomic Publishing Company. ISBN No. 1-56676-780-6.

MacDiarmid, A. G. (1997). Polyaniline and polypyrrole: Where are we headed? *Synthetic Metals, 84*(1997), 27–34.

Nagle, H. T., Schiffman, S. S., & Gutierrez-Osuna, R. (1998). The how and why of electronic noses. *IEEE Spectrum, 35*(9), 22–34.

Nakamoto, T. (2016). *Essential of machine olfaction and test* (p. 235). Wiley, Chapter Five. ISBN: 9781118768488.

Nguyen, D., & Yoon, H. (2016). *Recent advances in nanostructured conducting polymers: From synthesis to practical applications*. MDPI. Open Access. <https://www.mdpi.com/2073-4360/8/4/118/htm>.

Nose Office. (2003). *NOSE II—The second network on artificial olfactory sensing*. Tübingen: University of Tuebingen.

Ohloff, G. (1990). *Scents and fragrances*. Berlin: Springer-Verlag.

Persaud, K., & Dodd, G. (1982). *Nature, 299*, 352–355.

Razak, A. (2007). *Gas turbine combustion. Industrial gas turbines*. ScienceDirect. https://www.elsevier.com/books/industrial-gas-turbines/razak/978-1-84569-205-6.

Sberveglieri, D., (1999). Metal-oxide semicondiuctors. In: *ASTEQ technologies for sensors*.

Stetter, J. R., & Penrose, W. R. (2001). *The electrochemical nose*. Chicago, IL: Department of Biological, Chemical and Physical Sciences, Illinois Institute of Technology.

Vaefolomeev, S. (1999). Conducting polymer sensors. In: *ASTEQ technologies for sensors*.

Wackerly J. (2004). Conjugated polymers as fluorescence-based chemical sensors. <https://chemistry.illinois.edu/system/files/inline-files/06_Wackerly_Abstract.pdf>

Wilson, A. D., & Baietto, M. (2009). Review—Applications and advances in electronic-nose technologies. *Open Access—Sensors*, *9*, 5099–5148. Available from https://doi.org/10.3390/s90705099. ISSN 1424-8220 <http://www.mdpi.com/journal/sensors>.

# Comparison and validations

Comparison among various devices is one approach in providing needed validation for the output of machine olfaction device (MOD), that is in the form of accurate results via testing various samples using one or more than one application (see Section 2.6).

Also, the validation techniques in MOD are for the purpose of calculating the device error rate under normal as well as certain defined conditions. Normal conditions simply mean the environment and conditions the device was designed for while certain conditions may include a change of temperature, humidity, wind, and pressure.

Data Gaussianity (Gaussianity is a normal distribution curve, i.e. it forms the shape of a bell or called the bell curve) of the device should be verified, which may involve different methods, such as individual sensor values, using main components for linear projection of data (scatter plot), multivariate kurtosis and skewness, and other related methods.

## 2.1 Background

It is difficult to pinpoint the exact date of when and how the idea of designing a system which can mimic the human nose came about. However, the following dates and names provide a better idea on how MOD system progressed in modern times (for more details see the following Section 2.1.1).

### 2.1.1 Brief historical background

The following are the names that designed and developed the eNose in various ways from the mid of the 20th century up to around the end of it.

Hartman 1954
Moncrief 1961
Wilkens, Hatman 1964
Buck, Allen, Dalton 1965

Introduction to Machine Olfaction Devices.
DOI: https://doi.org/10.1016/B978-0-12-822420-5.00005-2
© 2022 Elsevier Inc. All rights reserved.

Dravieks, Trotter 1965
Persaud & Dodd 1982
Ikegami 1985
Persaud, Pelosi 1985
Nakamoto, Moriizumi 1989
Sundgren et al. 1990
Aishima 1991
Singh, Hines, Gardner 1996
Göpel et al. 1997

The above sequences of research work have led to the eventual conceptualization of the eNose.

Clearly the first recorded scientific attempt to use sensor arrays to emulate and understand mammalian olfaction was carried out by Persaud and Dodd in 1982, at the University of Manchester Institute of Science and Technology, United Kingdom. A device was built with an array of three metal-oxide gas sensors used to discriminate among 20 odorous substances. Using visual comparison for the ratios of the sensor responses, they obtained the pattern classification. It took 10 years before their idea produced the first commercial nose named "AromaScan."

The name electronic nose (or eNose) was used for the first time during 1988 and has come into common usage *as a generic term for an array of chemical gas sensors incorporated into an artificial olfaction device* (Stetter & Penrose, 2001) after the introduction of this title at a conference covering this field in Iceland 1991. From that point, the idea and the principles of the eNose has grown and developed into different fields across the globe.

Historically speaking, there are two different types of electronic noses:

1. Static odor delivery.
2. Mass-flow systems.

As the name suggests, the basic mechanism for the first type is that there is no odor flow but simply a flask containing the sensor arrays, with a fan at the top to distribute the flow within the flask. This type was the design of the first eNose in 1982. The second type, which is very popular now, is where the odor flows within the system. The most recent eNose designs are made in this way.

To complete this brief historical outlook concerning the eNose, it is informative to look at the basic schematic comparison between human and electronic noses summarized in the following points (Davide, Holmberg, & Lundstrom, 2001):

A. The human nose

To understand how we can smell certain odours, the process take place when type of compounds referred to as "odorants" cause stimulation to the receptors located at the roof of nasal cavity (olfactory sensory neurons) within the olfactory epithelium and connected to the brain.

The main parts of the olfactory system are *the olfactory receptors* (detecting airborne odor molecules), *olfactory bulb* (send olfactory information), and *olfactory cortex* (processing and perception of odor). As volatile compounds, that is odorants, which are hydrophobic, weigh less than 300 daltons (in molecular weight), they enable the odorant receptors in the human nose to distinguish around 1000 types of substances (Pearce, Schiffman, & Gardner, 2002) (Fig. 2.1). The following is summary of the characteristics of the human nose.

1. There are millions of self-generated receptors (approximately 10 million).
2. Selectivity classes (range from 10 to 100).
3. Initial reduction of number of signals ($\sim$1000 to 1).
4. Very adaptive.
5. Saturation can happen—operate for short periods of time.
6. Real-time signal interpretation.
7. Variety of odors can be identified.
8. Can detect some specific molecules but it cannot detect some simpler molecules.
9. Infection can take place.
10. Smell can be associated with various experience and memory.

Fig. 2.2 is a schematic illustration of the human olfactory system as it contains large number of olfactory receptor cells ($\gg$10 million) but with a limited amount of selectivity classes ($\sim$10–100). An odor produces a pattern of signals to the olfactory cortex via the mitral cells ($\sim$10,000). The brain interprets the signal pattern as a specific odor (Box 2.1).

B. The eNose
1. Approximately 5–100 chemical sensors manually replaced.

**Figure 2.1**
Schematic diagram showing the location of the olfactory bulb, olfactory receptors, and olfactory cortex.

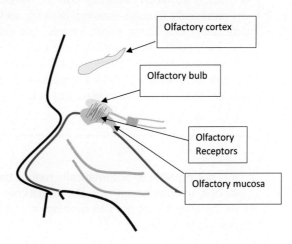

Olfactory cortex

Olfactory bulb

Olfactory Receptors

Olfactory mucosa

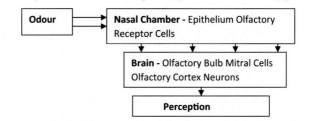

**Figure 2.2**
Scheme of the human olfactory system.

---

## Box 2.1 Human olfactory system.

The olfactory system in humans is still the subject of much debate. Simply put, what we smell is nothing more than stray molecules of substances, known as odorants, that waft through the air and settle into molecular receptors inside our nostrils. The odorants induce an electric signal in the receptors that is carried through neurons into the olfactory bulb, which sends signals into a brain structure known as the olfactory cortex, which passes them along until they eventually reach the hippocampus, a primitive brain structure associated with memory in mammals.

*(Taubes, 1996)*

The process of olfaction is initiated when volatile chemicals stimulate olfactory receptor neurons located on a relatively small patch of specialized epithelial tissue high in the nasal cavity. These sensory neurons have axons that travel as the olfactory nerve (cranial nerve I) to terminate in the olfactory bulb. In turn, the olfactory bulb projects more centrally and contributes inputs for higher cortical processing, which results in olfactory perception. Odorants reach the olfactory receptors in two ways: they can enter the nostrils during normal inhalation (orthonasal route) or travel from the back of the oral cavity to the olfactory receptors via the nasal pharynx (retronasal route). The perception of food flavor involves a combination of olfactory activation caused by odorous compounds released into the nasopharynx retronasally through chewing, drinking, and deglutition, and the blending of taste (salty, sour, bitter, sweet, umami) and oral somatosensory sensations (texture, heat, cold). Nasal blockage and swelling can prevent odors from entering the retronasal stream, resulting in a shift of flavor toward blandness. Reported taste loss is more typically a loss in the perception of food flavor due to loss of retronasal olfactory function than to a decrement in taste perception, per se.

*(Dalton et al., 2013)*

---

2. It is not possible to automatically reduce the number of signals to a particular one, in comparison with the human nose.
3. As the eNose continues to develop, it is possible that it may become adaptive in the future.

4. It is unlikely to become saturated and can work for long periods of time.
5. If pattern recognition hardware is provided within the device, then new real-time signal treatment can occur.
6. Unlike the human nose, the eNose needs to be trained for each application.
7. It can detect simple molecules and it cannot detect some complex molecules at a low concentration.
8. The eNose can get poisoned (sensors—loss of function).
9. It is possible for an eNose with multisensors to be associated with other functions/recognitions.

## 2.2 How does it work?

Several operation parameters are usually required for the eNose to function to maximum effect (Fig. 2.3). These can be:

1. Setting up the temperature for the sample incubation.
2. The size of the sample.
3. The rate of injection.
4. The quantity of injection.
5. The added solvent being used.
6. Flow rate.
7. Sensor type.
8. Sensor operational parameters.

The above are a few examples; there can be others as well.

As mentioned briefly earlier, the principle of eNose mainly rests with one or more (an array) of vapor-sensitive detector (sensor). Usually the detector is made up of certain types of sensitive materials whose characteristics/behaviors change in response to

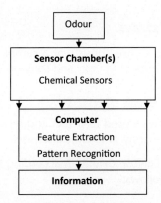

**Figure 2.3** Schematic diagram of an electronic nose (MOD). A limited number of chemical sensors (10–100) with partly overlapping selectivity profiles. A computer is used to extract the features from the sensor signals and to recognize the patterns belonging to a given odor or gas mixture.

absorbed or adsorbed molecules. As we measure the changes in each sensor, identification can be made for the unknown odor(s) by comparing it with the library data.

As we are scaling down (in size) the present MOD, the design parts of the new device must be simple and easy to configure, in particular the utilization of the MMOS (or other types of sensors, if used instead of MMOS). This therefore may include a detailed study of the materials that will be used. This will include the understanding, as well as the applications for some of the instruments/engineering methods required.

## 2.3 zNose

### 2.3.1 What is the zNose?

The zNose is a detection device; therefore it is more like a detector than an analytical device (e.g., eNose). What the above means is that zNose functions as gas chromatography (GC) function; however, the outputs measured in relation to time are similar to eNose. In comparison, the eNose is much more widely used in different fields compared with the zNose.

The zNose depends on its analysis of vapors, odors on the changes of quartz crystal vibrations with each element, that is sound. The principles used here are very much like those used in the Quartz Crystal Microbalance (QCM) approach. QCM and QCM-D (Quartz Crystal Microbalance with Dissipation) are types of materials where their mass changes at a nanoscale level as and when molecules interact with the surface and consequently mass changes occur.

### 2.3.1.1 Differences between zNose and eNose

Table 2.1 has been included for general information and in order to show the differences between zNose and eNose (Staples, 2000).

## 2.4 eNose

As a device, the eNose consists of a sample delivery system, chamber for the sensors, signal processing unit, and software.

### 2.4.1 What is an eNose?

An eNose is a device used to detect and recognize odors/vapors; in other words an MOD with an array of chemical sensors.

Alternatively, an eNose definition according to Gardner and Bartlett (1994), has been stated as follows:

> An electronic nose is an instrument which comprises an array of electronic chemical sensors with partial specificity and an appropriate

**Table 2.1** Comparison between zNose and eNose according to Edward J. Staples from "Electronic Sensor Technology."

| ZNose compared with eNose | | |
| --- | --- | --- |
| | ZNose | eNose |
| Speciation | Yes | No |
| EPA methods | Yes | No |
| Sensitivity | ppb | ppm |
| Speed | Seconds | Minutes |
| Intelligence | Human | Artificial |
| Precision | High | Low |
| Accuracy | High | Low |
| Stability | Months | Hours |
| Cal standard | Yes | No |
| Sensors | Hundreds | 4–32 |
| Olfactory images | Yes | No |

*Source*: Adapted and redrawn from Staples, E. J. (2000). The first quantitatively validated electronic nose for environmental testing of air, water, and soil. *ACS National*, 26–30.

pattern recognition system, capable of recognizing simple or complex odors.

The following are just examples of some of the areas where the eNose can be used:

Food in general and flavor analysis of foods in particular.
Petroleum—qualitative and quantitative analysis.
Vaginal smears, amniotic fluid detection.
Detection of explosives.
Classification and degradation studies of olive oils.
Development of a field odor detector for environmental applications.
Quality control applications in the automotive industry.
Discrimination between clean and contaminated cow teats in a milking system.
Cosmetic raw materials analysis.
Checking and the differentiation of various wine aromas.

The principles of eNose are that it uses "an array of sensors," whether in the form of different types of polymers or via the use of metal-oxide semiconductors, for which the same principle applies.

To conclude this section of the chapter, and as mentioned earlier, when molecules from any element are deposited on the surface of a sensor, the electrical conductivity usually changes as and when the surface expands, and the reverse produces the opposite results (Box 2.2).

---

## Box 2.2 Detection capability.

... a high quantity of water is sometimes present, so the headspace turns to be represented by a mixture of volatile organic compounds in an atmosphere containing a high percentage of relative humidity. To this reason, before testing an electronic nose to the aroma of a specific food it is important to verify the detection capability and the sensitivity of the sensors toward these main volatile compounds, mainly to distinguish the contribution to the response coming from the humidity and from the compounds themselves.

*(Taurino et al., 2003)*

---

# 2.5 SMart nose

## 2.5.1 What is SMart nose?

SMart nose is the commercial name for a mass spectrometry (Ms) device. This device allows the direct characterization of volatile organic components, from liquid and solid, using Ms techniques, that is to say, without separation of the headspace constituents. Aryballe, a digital olfaction company, located in Grenoble, France manufacture Smart Nose, such as NeOse Pro.

The training required for eNose will not be applicable on this device.

Additional details will be provided in the following chapters.

# 2.6 Machine olfaction device validation (standard analytical method)

## 2.6.1 Gas chromatography

GC is a very sensitive analyzing method for separating small amounts of liquid or gaseous mixtures and quantitating the amount for each separated component. It is used for volatile compounds to separate the sample into its various constituents and elements.

This method involves a sample being vaporized and injected onto the head of the chromatographic column. The sample is transported through the column by the flow of inert (gaseous) mobile phase. The column itself contains a liquid stationary phase, which is adsorbed onto the surface of an inert solid.

This method is used for volatile compounds to separate the sample into its various constituents and elements (Fig. 2.4).

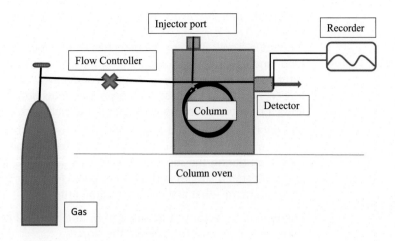

**Figure 2.4**
Schematic diagram of a gas chromatograph (GC).

## 2.6.2 Instrumental components

*Carrier gas*—It must have chemically inert gases, for example Nitrogen, Helium, Argon, and Carbon Dioxide. The gas chosen is mainly dependent on the type of detector.

*Sample (injection)*—The sample should not be too large, as slow injection of large samples causes band broadening and possible reduced resolution.

The temperature of the sample port can vary but is usually around 50°C or set a little above boiling point of the compounds under analysis.

*Columns*—Two types of columns:

a. *Packed* 1.5–10 m length with an internal diameter of 2–4 mm.
b. *Capillary* (also known as *open tubular*): Internal diameter about one-tenth of a millimeter with two types available: wall-coated open tubular or support-coated open tubular.

*Column temperature*—Column temperature must be controlled to the precise temperature required (within tenths of a degree). There should be a relation-ship between the boiling point of the sample and the *optimum* column temper-ature, as the column temperature is dependent on it. The column oven temperature is usually increased incrementally, or else long analysis times can result, with consequent band broadening.

*Detectors*—In GC various types of detectors can be used. Consequently, these different detectors can give different types of selectivity.

Non-selective *detector: These react to all compounds except the carrier gas.*

Selective *detector: React to a range of compounds with a common physical or chemical property.*

Specific *detector: React to a single chemical compound.*

A potential application for the detectors can be found mostly within the medical, food, and environmental sectors. Examples of detectors and their usage include: flame ionization (mass flow), thermal conductivity (concentration), electron capture (concentration), nitrogen phosphorus (mass flow), flame photometric (mass flow), photoionization (concentration), and Hall electrolytic conductivity.

The data (qualitative and quantitative) are usually obtained after the separation of the mixture by GC, using a detector. Detectors continuously generate electrical signals during the separation. A linear signal is generated after the detection of the separated species and will form peaks for every component, displayed as a chromatogram—that is signal versus time output.

The type of detector will be decided by the range of compounds for which the detector will be able to generate the signals.

### 2.6.2.1 Static headspace gas chromatography (HSGC)

To analyze volatile organic compounds, a simple technique called static headspace gas chromatography (SHGC) is usually used. SHGC techniques are very well-known globally for analyses of alcohols in blood and residual solvents in pharmaceutical products (Restek, 2000). There are other applications for SHGC as well; these can range from flavoring compounds in beverages and food products, to fragrances in perfumes and cosmetics.

We may get poor analytical results if samples contain nonvolatile material; for this reason sample preparation is important as it will help to extract the compounds being investigated from unwanted, nonvolatile substances. SHGC analysis can overcome this by directly sampling the volatile headspace from the container in which the sample is placed (Box 2.3).

---

## Box 2.3 Headspace/GC sample preparation

Samples for headspace/GC must be prepared in such a manner as to maximize the concentration of the volatile sample components in the headspace while minimizing unwanted contamination from other compounds in the sample matrix. Sample matrices such as biological samples, plastics, and cosmetics can contain high molecular weight, volatile material that can be transferred to the GC system. Water from the sample matrix also can cause problems by recondensing in the transfer line. Incomplete or inefficient transfer of high molecular weight compounds or water vapor from sample matrices can produce adsorptive areas in the transfer line or injection port

*(Continued)*

---

**Box 2.3 (Continued)**

that can lead to split peaks, tailing peaks, or irreproducible responses or retention times. To minimize matrix problems and prevent water condensation from aqueous samples, use a higher transfer line temperature ($\sim 125°C-150°C$).

*(RESTEK, 2000)*

### 2.6.3 Flame ionization detector

In a flame ionization detector (FID) the outflow is mixed with hydrogen and air and ignited. Ions and electrons are produced as a result of burning organic compounds; these can conduct electricity via the flame. There will be a large electrical potential applied at the tip of the burner, and at the same time there will be a collector electrode located close to the flame. The obtained current measure results from the "pyrolysis" of the organic material.

The positive aspects of FID organic compound detector are that it generates low noise and is simple to use. It also has a high sensitivity and a good response range. However, using a device such as this one can result in the destruction of the analyte.

### 2.6.4 Mass spectrometry

The Ms can identify the type of materials in the sample by electrically charging the specimen molecules. These molecules will be accelerated via a magnetic field. Consequently, the molecules will be broken into charged fragments and then the different charges can be detected. The mass of each fragment will be displayed in the form of a spectral plot, and then the compound's mass spectrum can be used for qualitative identification. The process here is that the fragment masses of the molecules are used to piece together the mass of the original molecule.

The analyzing work therefore is that from the molecular mass and the mass of the fragments, reference data are compared to find out the identity of the sample (Douglas, undated). This is possible because each substance's mass spectrum is unique, as long as the parent mass correctly fits the output or vice versa.

#### 2.6.4.1 Process description

The basic description of Ms is that it contains a sample inlet, an ionization source, a molecule accelerator, and a detector. Of course, there are many variations of Ms but, to be brief, a look at the conventional and basic Ms will be sufficient.

To make sure that the sample needed to be analyzed will stay as a gas, the sample inlet is kept at a temperature of 400°C, mainly because in Ms analysis there is a need for a pure gaseous sample. However, the above temperature does not have to be always as high as 400°C; since the inlet is at low pressure, some organic molecules stay volatilized at a temperature as low as 250°C.

The output is produced in the shape of array of peaks on a chart, called the mass spectrum. Every "peak" is equivalent to the value of a fragment mass. The more fragments detected with one particular mass, the higher the peak will be (Douglas, undated).

*Output analysis*—Under certain controlled conditions, each substance has a characteristic mass spectrum. That means it is possible to identify a specimen by comparing the specimen's mass spectrum with known compounds. In measuring relative intensities of the mass spectra, only then quantitative analysis possible.

For the unfragmented molecule sample, a mass spectrum usually displays a peak. As mentioned above, it is called parent mass and it is the biggest detected mass.

The mass of the molecule is revealed by the parent mass; however, various other peaks reveal the molecule's structure.

Probably the hardest part of Ms analysis is looking at fragmentation, finding the parent peak, and then the molecular mass of the sample. Obviously, in various spectral analyses, computers will be needed for this purpose.

Finally, not all compounds give a peak for the entire molecule. This is called the molecular ion, not the parent mass (Box 2.4).

---

## Box 2.4 Ion free path

All mass spectrometers must function under high vacuum (low pressure). This is necessary to allow ions to reach the detector without undergoing collisions with other gaseous molecules. Indeed, collisions would produce a deviation of the trajectory and the ion would lose its charge against the walls of the instrument. On the other hand, ion—molecule collisions could produce unwanted reactions and hence increase the complexity of the spectrum.

*Hoffman and Stroobant (2007)*

---

## 2.7 Software

In order for the MOD to be able to provide meaningful data, as a result of sensor's reaction to the sample, the software employed for the device uses a series of algorithms and chemo-metric methods.

Signal processing, pattern recognition, and data analysis can be controlled and handled and maintained within one computer program.

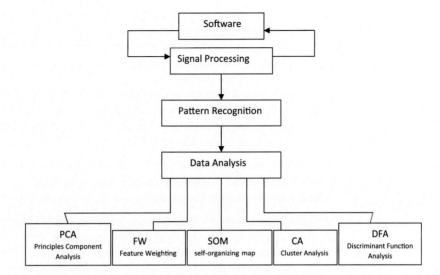

**Figure 2.5**
Diagram illustrating the stages of the processing after sample analyzing.

Fig. 2.5 shows the software stages with various methods of data analysis.

There are number of software programs that can be used to operate and control various MODs.

The software's function here is mainly to control the device components, as well as to provide reading and pattern recognition.

A program can also be embedded in the chip (firmware), usually described as an embedded system. An embedded system has been designed specifically to meet different requirements than a general purpose computer system. Some examples of where embedded systems can be found are "handheld calculators," "mobile phones," "computer printers," "multifunction wristwatches," and in "medical equipment."

The benefits of an embedded system are that it can maintain 100% reliability, despite constantly being "on" over a very long period of time (many years) (Bernstein et al., 1998).

There are many types of embedded software architectures, for example, "The Control Loop," "Nonpreemptive Multitasking," "Preemptive timers," "Preemptive tasks," "Office-style Operating Systems," and "Exotic Custom Operating Systems."

## 2.8 Data processing

What we get from the sensor (i.e. the information) can be described as a stream of data related to certain functions (e.g., current resistance against time). The eNose therefore must convert the information (data) into a usable format. The critical part in the function of the MOD therefore is the analyzing of the signals generated by the sensor. However, in order to improve on MOD signal, that is raw MOD signals, a noise filtering method can be selected to remove or minimize the noise. There are a number of noise filtering methods that can be considered for the above purpose, such as wavelet transform (mathematical approach dividing a function/continuous-time signal into various components), binomial smoothing (this method is based on the binomial coefficients for implementing Gaussian filtering), and moving average (a calculating method for the purpose of analyzing data points via the making of a series of averages of various subsets of all data set).

Methods for data presentation techniques such as employing what has been termed "scaled polar plots," "difference plots," "difference rings" plus various other methods, are simply to allow for qualitative sample-to-sample differentiation.

Artificial Neural Network (ANN) is another type of data processing technique that can be employed. This method has an advantage over other methods simply because ANNs are self-learning so that more data processed for the ANN means better detecting results. So the more the MOD/eNose is being used, the more learning takes place and more data stored in the computer memory. Consequently, the easier and quicker detection can be achieved as the device is used more often.

More details concerning the above topic will be discussed further in the following pages.

## 2.8.1 Using the sensor data

There is a characteristic value (or values) for the sensor connected to each exposed volatile gas (vapor).

The number of values ($n$) obtained from the sensor will be considered as a vector in $n$-dimension in a sensor space.

These vectors from different odor exposures will be compared, for example, by either the angle between the two vectors or the distance (Euclidean) between their termini.

Using standard pattern classification algorithms, the vectors will be processed (Fisher's Linear Discriminant) by taking sets of them (vectors and category tags) in order to identify groups of identical vapor. In the process, it will learn to correctly assign unknown exposed materials by comparison (Box 2.5).

---

### Box 2.5 Data acquisition

Data acquisition is the first step for data analysis; sensors collect the data and convert it into an electrical signal pattern that is more suitable for computer analysis. This step often causes the difficulty in the classification problem as the

*(Continued)*

---

> ## Box 2.5 (Continued)
> characteristics and limitations of the transducer may limit or distort the available information. The output is a pattern vector, in pattern space. The pattern vector is passed into the second stage, the feature extractor. Feature extraction is the use of one or more transformations of the input features to produce new salient features. Feature extraction may be regarded as a dimensionality reduction process; data are converted from pattern space into feature space. Features should be easily evaluated; there are two kinds, the first have a clear physical meaning, the second have not and are called mapping features.
>
> Scott, James, & Ali, 2006

## 2.8.2 Signals and reprocessing

The MOD is basically a device containing nonspecific sensor(s) alongside a pattern recognition system; therefore the data processing components can range from a powerful desktop to a limited computational portable system, depending on the final requirements needed and the actual size of the device desired.

A small portable device may need *embedded pattern analysis software*; therefore the complexity of training algorithms in many cases is much higher than during the operation stage. Therefore some algorithms should be trained off-line using a host computer, as the embedded software may not be sufficient for the type of operations for which we are intending to use the device. The process therefore from the sensor response through the final classification for the MOD can be summarized in the following points:

1. Sensor response (raw measurement).
2. Preprocessing.
3. Normalized measurements.
4. Feature extraction.
5. Classification.
6. Vapor/odor class.

However, in order to get accurate results, the MOD sensor(s) should be sensitive for any minute changes, which may take place around it. The sensor response will be amplified, and the data processing will interpret the final result. One of the main purposes of the preprocessing is to:

1. reduce noise;
2. remove drift;
3. translate the data.

The aim of the "preprocessing" in the MOD is simply to choose a number of values (parameters), which represent the sensor response. This is an important procedure as it may affect the performance of the following modules in the pattern analysis system while at the same time it is part of the work for the signal preprocessing to compensate for sensor drift and preparing the feature vector for further processing (Gutierrez-Osuna, 2002) (Fig. 2.6) (Table 2.2).

There are three steps of signal preprocessing:

Stage 1 "Baseline manipulation" is connection to the constant signal value coming from the sensor. The baseline signal therefore is the response of the sensor (e.g., MOS sensor) to air/ambient environment.

There are three methods used for this stage, which are *difference, relative, and fractional.*

> *Difference:* The baseline is subtracted from the sensor response.
> *Relative:* Divide the response by the baseline.
> *Fractional:* Subtract and divide by the baseline.

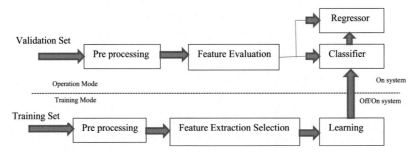

**Figure 2.6** Standard architecture of signal processing for eNose. The validation (normal mode) is generally computed in the device itself; however, training is mostly done using a computer. *Source: Redrawn from Perera, A., Sundic T., Pardo A., Gutierrez-Osuna R., & Marco S. (2002). A portable electronic nose based on embedded PC technology and GNU/Linux: Hardware, software and applications.* IEEE Sensors Journal, 2(3), 235.

**Table 2.2** Preprocessing techniques review.

|  | Technique | Sensor type |
|---|---|---|
| Baseline manipulation | Difference | QMB, MISFET |
|  | Relative | MOS |
|  | Fractional | MOS, CP |
| Transient compression | Subsampling | MOS |
|  | Parameter extraction | MISFET |
|  | Model fitting | MOS, CP, QMB |
| Normalization | Sensor, Vector, Auto-scale | MOS, CP |

*Source*: Adapted from Gutierrez-Osuna, R. (2002). Pattern analysis for machine olfaction: A review. *IEEE Sensors Journal, 2*(3), 189.

In Stage 2, "Compression," there is a need for the calculation of various descriptors from the resulting response curve. Usually, the sensor steady-state reaction is used.

In Stage 3, "Normalization," there are procedures that can be used in order to reduce what has been described as the bias of different descriptor scales related to multivariate statistical techniques. The method used will rely mainly on the pattern recognition type needed. Therefore a Feedforward Neural Network could be more suitable for "Range Scaling" and similarly, "Auto-Scaling" is more appropriate for Principal Components Analysis (PCA).

## 2.8.3 Pattern recognition

Multivariate statistical and neural networks methods, based on instantaneous sensor responses, are usually used for pattern recognition. The above two methods need many tests for each sample in order to build the predictive model, which can sometimes be time-consuming.

Multivariate statistical methods, such as "PCA," are usually used for the MOD response, mainly because they are not affected by collinearity, which can be a problem in sensor arrays.

These analyses of the interrelationships between various variables are described in terms of their common underlying dimensions, known as principal components. Pattern recognition therefore can be an important part in the make-up of the eNose as each highly sensitive sensor can partially overlap others in the sensors array. Therefore, the pattern recognition techniques are necessary for analyte discrimination (Box 2.6). Possible methods used in this respect can be "PCA," "ANN," "Discriminant Analysis (DA)," "Cluster Analysis," and "Feature Weighting."

---

### Box 2.6 Data treatment

The combination of responses generated by the array of sensors is not specific for a certain compound, but may be related to certain characteristics of the samples by means of pattern recognition techniques. Four sequential steps are followed: signal preprocessing, dimensionality reduction, prediction, and validation. ... The signal preprocessing includes drift compensation, scaling, and the extraction of representative parameters. The dimensionality reduction step projects the initial feature onto a lower dimensional space.... This is usually carried out by means of a nonsupervised technique, such as PCA. PCA can be used to discriminate between samples with different organoleptic characteristics. The resulting low-dimensional feature vector is used to solve prediction problems, such as classification or regression.

*Rodríguez-Méndez et al. (2016)*

---

Brief additional materials related to the above are provided below. However, in addition to this, the following chapter(s) will provide further details for these methods.

## Types of ANN

Modular neural networks.
Feedforward neural network—artificial neuron.
Radial basis function neural network.
Kohonen self-organizing neural network.
Recurrent neural network.
Convolutional neural network.
Long-/short-term memory.

## Types of DA

Multiple DA
Fisher's Linear Discriminant Analysis (LDA)
K-nearest neighbors (KNN) DA
Discriminant factor analysis
Canonical DA

## Types of clustering

Connectivity-based clustering (hierarchical clustering).
Centroids-based clustering (partitioning methods).
Distribution-based clustering.
Density-based clustering (model-based methods).
Fuzzy clustering.
Constraint-based (supervised clustering).

To clarify the above, under linear classification methods, we have "LDA," "KNN," as well as "support vector machine (SVM)," which can train and identify high dimensional samples in gas recognition.

According to Faleh, Bedoui, & Kachouri (2020) when it comes to comparison of the main algorithms used for gas detection, LDA is found to be a good method when it comes to interpretation, ease of implementation, execution time, and very good for classification quality. KNN, for the same above parameters, its performance is very good for the first four, and good for the classification quality. SVM is good for execution time and very good for the rest of the above parameters. ANN is very good for classification quality, while it is poor when it comes to interpretation, ease of implementation, and very poor for execution time.

There is also the "Independent Component Analysis," which is not mentioned above, as another method that is able to decompose a data matrix into components each independent from the others that can be applied to the MOD data sets.

Finally, what has been termed as "extreme learning machine (ELM)" is another fast-learning algorithm in neural network, that is part of a hidden single layer for feedforward. The ELM's function is to determine output's weights after randomly generating the hidden node parameters (Peng et al., 2016).

Additional details related to algorithms/data can be found in Chapters 3 and 9.

# References

Bernstein, P., Brodie, M., Ceri, S., DeWitt, D., Franklin, M., Garcia-Molina, H., ... Ullman, J. (1998). The Asilomar report on database research. *SIGMOD Record, 27*(4), 74−80.

Dalton, P., Doty, R., Murphy, C., Frank, R., Hoffman, H., Maute, C., ... Slotkin, J. (2013). Olfactory assessment using the NIH toolbox. *Neurology, 80*(11 Suppl 3), S32−S36. < https://www.ncbi.nlm.nih.gov/pmc/articles/PMC3662337/ >.

Davide, F., Holmberg, M., & Lundstrom, I. (2001). 12 virtual olfactory interfaces: Electronic noses and olfactory displays. *Communications through virtual technology: Identity community and technology in the Internet Age* (pp. 193−220). IOS Press.

Douglas, F. (undated) GC/MS analysis. Scientific Testimony—An Online Journal patentattorney@law.com or <http://sites.netscape.net/dougfrm>.

Faleh, R., Bedoui, S., & Kachouri, A. (2020). Review on smart electronic nose coupled with artificial intelligence for air quality monitoring. *Advances in Science, Technology and Engineering Systems Journal, Vol. 5*(2), 739−747. < https://www.astesj.com/publications/ASTESJ_050292.pdf >.

Gutierrez-Osuna, R. (2002). Pattern analysis for machine olfaction: A review. *IEEE Sensors Journal, 2*(3), 189.

Hoffman, E., & Stroobant, V. (2007). *Mass spectrometry: Principles and applications* (3rd ed., p. 10) John Wiley & Sons Ltd, ISBN 978-0-470-03310-4 (HB) ISBN 978-0-470-03311-1 (PB).

Pearce, T., Schiffman, S., & Gardner, J. (2002). *Handbook of machine olfaction.* Wiley. Print ISBN:9783527303588 |Online ISBN:9783527601592. < https://onlinelibrary.wiley.com/doi/book/10.1002/3527601597 >.

Peng, C., Yan, J., Duan, S., Wang, L., Jia, P., & Zhang, S. (2016). Enhancing electronic nose performance based on a novel QPSO-KELM model. *Sensors, 16,* 520. Available from https://doi.org/10.3390/s16040520.

Perera, A., Sundic, T., Pardo, A., Gutierrez-Osuna, R., & Marco, S. (2002). A portable electronic nose based on embedded PC technology and GNU/Linux: Hardware, software and applications. *IEEE Sensors Journal, 2*(3), 235.

Persaud, K., & Dodd, G. (1982). Analysis of discrimination mechanisms in the mammalian olfactory system using a model nose. *Nature, 299,* 352−355. Available from https://doi.org/10.1038/299352a0.

Restek. *A technical guide for static headspace analysis using gc technical guide—Lit. Cat.# 59895A.* (2000). <https://www.restek.com/pdfs/59895B.pdf or http://www.restek.com>.

Rodríguez-Méndez, M. L., et al. (2016). Electronic noses and tongues in wine industry. *Frontiers in Bioengineering and Biotechnology.* Available from https://doi.org/10.3389/fbioe.2016.00081.

Staples, E. J. (2000). The first quantitatively validated electronic nose for environmental testing of air, water, and soil. *ACS National,* 26−30.

Stetter, J. R., & Penrose, W. R. (2001). *The electrochemical nose.* Chicago, IL: Department of Biological, Chemical and Physical Sciences, Illinois Institute of Technology.

Scott M.S., James D., Ali Z. (2006) Data analysis for electronic nose systems <https://www.researchgate.net/publication/226437091_Data_analysis_for_electronic_nose_systems>.

Taubes, G. (1996). The electronic nose. *Discover* (September Issue). <http://discovermagazine.com/1996/sep/theelectronicnos867>.

Taurino, A. M., et al. (2003). Quantitative and qualitative analysis of VOCs mixtures by means of a microsensors array and different evaluation methods. *Science Direct, Sensors and Actuators B, 93* (2003), 117–125. < https://pdfs.semanticscholar.org/0fd4/9dd9804e84369286234f365ec63-d43afbb7c.pdf > .

# MOD data and data analysis

Data provided from the sensors, as well as the type of information required (for a sample), will help in choosing a particular method for analyzing the final outputs. Having said that, it is important to have certain statistical knowledge as it will enable us to provide the right approach in collecting data and analyzing it. In this way, we can obtain accurate results needed for a particular solution, whether the outcome regarding a decision or a prediction, therefore, familiarity with statistical methods is essential.

Statistics offers a good approach in the way data are collected, summarized, and presented as a set of data that can be used as an outcome in the form of information needed. Variable or variables are used, which can be in the form of a sample. This means by obtaining the observed value of a variable, data are obtained.

From statistical approach, data are categorized under three headlines:

1. Interval data—are quantitative data or real numbers, for example, length, prices, heights.
2. Nominal data—are qualitative or categorical data, that is, related to names or categories, for example, male, single, family.
3. Ordinal data—are the values contained in data but with specific order, which is a categorical data, such as the level obtained in university degrees can be: pass, lower second class, upper second class, first class.

When it comes to the types of statistical tests, there are variety of them, such as Anova test, *t*-test, one sample median test, *Z*-test, chi-square test, binomial test, one sample median test, and parametric tests.

For processing and analyzing data, MODs use various computer software programs.

The software allows the user to load data files created by the MOD for visualization, such as via the Principal Components Analysis "PCA" (see Section 3.1.2.1).

**Introduction to Machine Olfaction Devices.**
**DOI: https://doi.org/10.1016/B978-0-12-822420-5.00009-X**

There are three analysis approaches used for MOD, such as graphical analyses (e.g., bar chart, profile, polar, and offset polar plots), multivariate data analyses—multivariate data analysis (MDA) (e.g., "PCA," canonical discriminate analysis "CDA," featured within "featured weighting (FW)" and cluster analysis "CA"), and network analyses (e.g., artificial neural network "ANN" and radial basis function).

Separate file is created by MOD for each sample that has data concerning configuration and measurement parameters related to exposure time, as well as data related to the sensor response for the user updating, such as the percentage deviation from the original sensor baseline.

Regarding PCA, the input matrix is constructed from a number of features extracted from each of the sensors in the respective sensor modules. In general, data acquisition can be summarized in the following steps:

Data collected via the sensors.

Data converted into an electrical signal, into pattern, that is, an output of a pattern vector.

Pattern vector → feature extractor → salient features

Computer's pattern analysis.

Other approach in data processing is the use of artificial intelligence (AI), such as machine learning or deep learning (see Chapter 9: Tests and Training—Section 9.6.1).

If the reader is not familiar with some of the basic statistical terms commonly used, the following is a brief reminder.

Basic statistical terms that will be referred to, either during general data discussion or when using one of the analytical methods mentioned above, can be a matrix, which is commonly referred to as a two-dimensional array of numbers of formulas. The word vector, such as a vector matrix, is a one row or one column. A one column is referred as a column vector, in the same way as a one row is referred to as a row vector. The number of rows multiplied by the number of columns matrix is referred to as dimension of matrix. When the number of columns and rows are the same it is referred to as square matrix; therefore a matrix dimension is the number of rows multiplied by the number of columns.

## 3.1 Data analysis

Methods of data analysis and their properties (Table 3.1) may vary, as it depends on the type of data and what we are looking for, and more importantly, the characteristic of the data analysis method used.

**Table 3.1** Methods of data analysis and their properties (Karakaya, Ulucan, & Turkan, 2020).

| Algorithm | Advantages | Disadvantages |
|---|---|---|
| PCA | Reduce the data dimensionality | High computational time with large amount of data |
| | Provide set of uncorrelated components | |
| | Measure probability estimations of high-dimensional data | |
| Discriminant analysis | Easy to use | High computational time for training |
| | Fast in classification applications | Complex matrix operations |
| | Linear decision boundary | Gaussian assumption |
| Regression | Easy to use | Outliers can cause problems |
| | Usage in several cases | Overfitting occurs |
| | Great with small amount of data | |
| ANNs | Can work with incomplete knowledge | Hardware-dependent computational time |
| | Fault tolerance is high | Hard to find optimal network structure |
| | Parallel processing | Overfitting may occur |
| | Once trained, predictions are fast | |
| | Effective with the large amount of data | |
| | For both regression and classification | |
| SVMs | Effective in high-dimensional spaces | High computational time with large amount of data |
| | Relatively memory efficient | Noisy data can cause overlapping classes |
| | Fast in both binary and multiclass classification | |
| | Works great with nonlinear data | |
| | Effective even in the cases when the number of dimensions is greater than the number of data samples | |

*Source*: From Karakaya, D., Ulucan, O., & Turkan, M. (2020). Electronic nose and its applications: A survey. *International Journal of Automation and Computing 17*(2), 179–209. https://doi.org/10.1007/s11633-019-1212-9

As an example, the number of features that have been extracted from the sensors, such as from PCA input matrix, mean that the curve parameters extracted may include:

1. The maximum deviation from the sensor baseline attained during the analytical procedure represented in the maximum response.
2. The maximum positive difference between two successive measurement values in the sensor response, represented by maximum slope.
3. The mean of the gradient of logarithm of the decay following the maximum response, that is, mean gradient of the logarithm.

For this reason, selecting an appropriate data analysis method is an important part in obtaining what we are intending to find out or the accuracy of the prediction.

As mentioned above, there are number of methods for data analysis which can be used, such as graphical analyses, discriminant function analysis, and CA (Box 3.1).

---

## Box 3.1 Types of cluster analysis.

There are three primary methods used to perform CA:

Hierarchical cluster: This is the most common method of clustering. It creates a series of models with cluster solutions from 1 (all cases in one cluster) to *n* (each case is an individual cluster). This approach also works with variables instead of cases. Hierarchical clustering can group variables together in a manner similar to factor analysis. Finally, hierarchical cluster analysis (HCA) can handle nominal, ordinal, and scale data. But remember not to mix different levels of measurement into your study. K-means cluster: This method is used to quickly cluster large data sets. Here, researchers define the number of clusters prior to performing the actual study. This approach is useful when testing different models with a different assumed number of clusters. Two-step cluster: This method uses a cluster algorithm to identify groupings by performing pre-clustering first, and then performing hierarchical methods. Two-step clustering is best for handling larger data sets that would otherwise take too long a time to calculate with strictly hierarchical methods. Essentially, two-step CA is a combination of hierarchical and k-means CA. It can handle both scale and ordinal data, and it automatically selects the number of clusters.

*Foley, 2018.*

---

### 3.1.1 Graphical analysis

Graphical analysis (GS) as a data evaluation tool is a statistical method, that is, an approach to solve a problem in the form of analyzing data via graphs techniques, such as histograms, data tables, Fast Fourier Transform, scatter plots, Sammon mapping, bar charts, dotplots. By doing so, GS provides the mechanism to understand pattern and correlation of parameters. GS is a quick method in collecting data samples, which can provide more accurate data outcome. There are two approaches in data analysis, the first one is descriptive statistics, which provide data summary obtained from the sample via the use of index, for example, standard deviation. The second method is inferential statistic, which deducts the outcome from data subject to random variation. Few examples of GS are presented below.

### 3.1.1.1 Bar charts

This is one of the simplest ways to obtain a visual response from the sensors, that is, in the form of a pattern. The pattern usually is in the form of height value of the displayed bars. The pattern obtained is the way the result or the composition of the sample or samples will be provided.

### 3.1.1.2 Polar plots (radar plots)

The plot provided is a multivariate two-dimensional data usually in the form of axis features that originate from the center in straight lines. These lines are with equal angles and the magnitude of the feature showed by the joined straight lines (see Fig. 3.1).

### 3.1.1.3 Hierarchical cluster analysis

This method is to rely on separating data into groups. These group of data are organized on the basis of similarity or close to similarity within each group. After creating these groups, the approach is to find similarity between these groups and connecting them together. In this way, a dendrogram will be created. The dendrogram presents a hierarchical relationship between the above groups. In this way, the dendrogram will help in allocating data into groups (see Fig. 3.2).

Finally, in choosing the correct graph in analyzing data, the following summary may help.

For attribute data, the use of bar charts, Pareto charts, and pie charts is mainly to compare between groups, while dotplots is to determine the shape of the data. On the other hand, time series plots is for process stability. The scatter plots is to examine the relationships among variables.

For continuous data, dotplots and histograms are for finding out the shape. Individual value plots and boxplots are for finding the differences between groups. Time series plots is for knowing process stability, scatter plots is to find

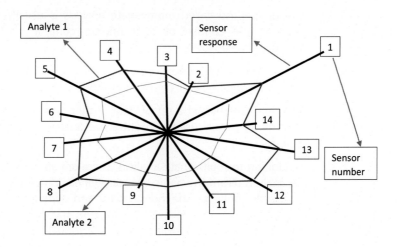

**Figure 3.1**
Illustration of sensor response data using polar plots.

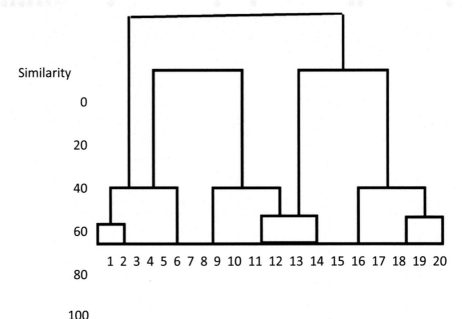

**Figure 3.2**
Illustration of HCA
data clustering in
the form of
dendrogram. *HCA,
hierarchical cluster
analysis.*

out the relationship between variables and lastly, multivari chart is to find out the source of variation.

### 3.1.2 Multivariate data analysis

As has been previously mentioned, multivariate data analyses cover "PCA," "CDA," "FW," "Feature Selection (FS)," and "CA."

When we have several variables, that is, complex sets of data, then a statistical tool or method called multivariate analysis is usually used in order to make sense of the context of their unclear content by measuring relationships between the variables.

Multivariate analysis therefore covers the evaluation and collection of data to help in explaining what the relationship of different variables from this data is.

Analyzing variables in a set of data, if there are independent observations, the assumption usually used in multivariate analysis is that the relationship should be linear.

In addition to multivariate analysis, there are also univariate analysis and bivariate analysis. From the name you can tell each method is related to the number of variables between the independent and dependent variables. The above will help to create a model by first analyzing the correlation and directionality of the set of data, which will provide the model (or fitting the line). Lastly, the above will help us to evaluate and validate the final result (outcome/usefulness) of the model (Box 3.2).

---

## Box 3.2 Multivariate data analysis techniques.

There are two categories of multivariate techniques, each pursuing a different type of relationship in the data: dependence and interdependence. Dependence relates to cause – effect situations and tries to see if one set of variables can describe or predict the values of the other one. Interdependence refers to structural intercorrelation and aims to understand the underlying patterns of the data. There are several multivariate models capable of finding those relationships, and many factors distinguish them. One of the primary factors that must be taken into account when choosing a technique is the nature of the data variables: they can be metric or nonmetric. Metric data variables are always of the numeric type and represent information that can be measured by some scale. Examples include age (20 years), temperature (25°C), and profit (United States $2000). The number specifies the magnitude of the value on a given scale. Nonmetric variables categorize the data, but do not specify its magnitude. Examples include an operational system (Windows, Linux, macOS) and house size (small, medium, large). The list of options that a nonmetric variable can assume is called levels or categories. Even when the levels have an inherent order (e.g., a large house is bigger than a small house), it is still a nonmetric variable because there is not any magnitude associated (the variable does not say how bigger is the house). Note that nonmetric variables can also be numeric when it is not attached to any scale, such as a variable that dictates the ID number of objects. Most multivariate techniques perform computations that need numbers as inputs, so how can a technique work with nonmetric data? The answer is that a nonmetric variable can become a dichotomic metric variable. In this conversion, each level becomes a new metric variable that can only have the values 0 (as false) or 1 (as true). For instance, consider a nonmetric variable that classifies the *color* of a product with the levels: black, white, and gray. The variable can be replaced with two new ones: *isColorBlack and isColorWhite*. If a product is black, then it assumes the values 1 and 0, respectively, and if it is white, the values 0 and 1, respectively. There is no need for a variable for gray products because they can assume the values 0 and 0: if they are not white nor black, they can only be gray.

*Divino Rodrigo (undated).*

---

The multivariate analysis historical background is briefly listed in the following points (Rao, 1983).

1. Sampling distribution paper by R. A. Fisher, the first step in multivariate analysis, 1915. Followed by the same author with a book titled "Statistical Methods for Research Workers" published in 1925.
2. The presentation of "The precise distribution of the sample covariance matrix of the multivariate normal population" by Wishart, 1928, which was an extension to Fisher's work.

3. Hotelling introduced his T2 statistic in 1931 providing the concepts of PCA in 1933 and canonical correlations in 1936.
4. The representation of D2 (the ratio of two independent chi-square variables) presented by Bowker, 1966.

### 3.1.2.1 Principal components analysis

PCA is a method used to reduce a large set of data into much smaller set, while keeping important and sufficient information needed from the original large data set, that is, dimensionally reduction technique. It can be compared to the olfactory receptor neuron in the human olfactory system convergence to the glomeruli (Persaud, 2016).

Dimensionally reduction techniques can be divided into FS and feature extraction. The purpose of FS and feature extraction is to reduce the number of features in a data set; however, FS is to select and exclude certain features with no change to the data, that is, keeps a subset of the original features, while feature extraction is to reduce dimensionally the data selected from the FS, that is, creates new set of data. Therefore PCA is a method used on large data sets changing it into much a smaller set while keeping sufficient information needed, that is, a linear data reduction technique method (PCA approach is to reduce multivariate data into two dimensions).

A set of uncorrelated new variables are found in PCA. The variance of this set usually decreases, and they are called the principal components (PCs) (Fig. 3.3).

The data obtained are usually represented in a multidimensional space, whose dimension is equal to the number of sensors in the array. A single measure is a $n$-dimensional vector.

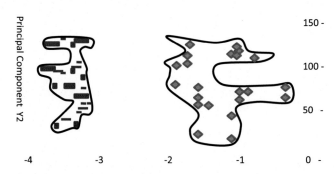

**Figure 3.3** A representative PCA plot in which multivariate data were reduced to two dimensions using PCA. *PCA*, Principal component analysis. *Adapted and redrawn from Griffin, M. (undated).* Electronic noses: Multi-sensor arrays. *Davidson, NC: Dept. of Chemistry, Davidson College.*

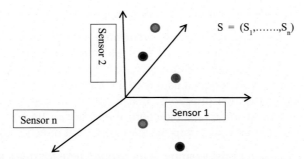

**Figure 3.4** Electronic nose data in multidimensional space representation. *The Sensor Space adapted and redrawn from Sberveglieri, D. (1999). Metal-oxide semiconductors. In: ASTEQ technologies for sensors (Sberveglieri, 1999) and Vaefolomeev, S. (1999). Conducting polymer sensors. In: ASTEQ technologies for sensors (Vaefolomeev, 1999).*

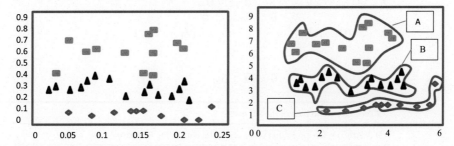

**Figure 3.5** Three odor analytes "scattergram"(left), the three odor analytes separated into three different classes "feature weighting" (right). *Adapted and redrawn from Griffin, M. (undated). Electronic noses: Multi-sensor arrays. Davidson, NC: Dept. of Chemistry, Davidson College.*

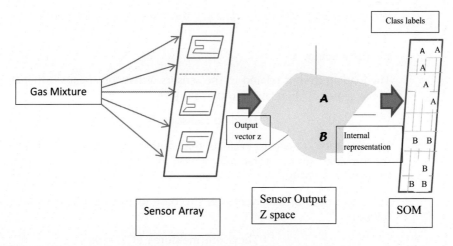

**Figure 3.6** eNose data analysis using SOM method. *SOM, Self-organizing map. Adapted and redrawn from Sberveglieri, D. (1999). Metal-oxide semiconductors. In: ASTEQ technologies for sensors and Vaefolomeev, S. (1999). Conducting polymer sensors. In: ASTEQ technologies for sensors.*

So how does PCA work? The following points will illustrate further (Brems, 2017):

1. Calculating how variables are related to each other in a matrix.
2. Dividing the matrix into two separate components: direction and magnitude. These two components are combinations of the original variables.
3. Aligning the original data with the above two components.
4. From the two directions, that is, direction and magnitude, by identifying the most important direction, the data can be compressed into a smaller space, that is, we drop the least important direction(s).

The following brief example is mentioned here for quick illustration.

Let **X** be the data matrix, the singular value decomposition of **X**:

$$X_{tm} = U_{tm}S_{mm}V_{mm} = S_1 \quad \begin{matrix} 0 \ldots \ldots 0 \\ 0 \quad S_2 \ldots \ldots 0 \\ (U, \ldots, U_m). \quad \cdot \quad \cdot \quad \cdot \quad \cdot \quad . \quad (V_1, \ldots V_m) \\ \cdot \quad \cdot \quad \cdot \quad \cdot \\ \cdot \quad \cdot \quad \cdot \quad \cdot \\ 0 \quad 0 \ldots S_m \end{matrix}$$

| |
|---|
| **V** (Vectors) = The base of the PC space |
| **U** (Vectors) = The projection of experimental data in the PC space |
| **S** (Scalars) = The singular values *(i.e., they are considered as a measure of the particular contribution to the systematic variance of the respective PC)* (Sberveglieri, 1999; Vaefolomeev, 1999) |

When we apply PCA to the data set, the result is two quantities:

a. The score.
b. The loading.

The loading itself is related to the sensors. What this means is that the loading of the sensor "i" in relation to the "PC j" is:

$$\text{Loading (sensor i, PC j)} = sj.vij.$$

Usually, loading helps to evaluate the contribution, which every individual sensor provides to the total information of the data set (Sberveglieri, 1999; Vaefolomeev, 1999).

The coordinate of data vector in the base of the PCs, that is, the score is as follows:

$$\text{The Score (measure i, PC j)} = uij$$

The result of analysis for the MOD in the form of a visual image, that is, the score plot—is usually limited to the most significant PCs.

PCA historical background

1. Deriving the singular value decomposition (SVD) by Beltrami (1873) and Jordan (1874).
2. The SVD used in a two-way analysis by Fisher and Mackenzie (1923).
3. The first technique presented as PCA was by Karl Pearson (1901).
4. Hotelling's further work on PCA in a paper (1933).
5. The asymptotic sampling distributions of the coefficients and variances of PCA by Girshick (1939).

PCA is also referred to in various literature as the Karhunen – Loève transformation, the Hotelling transformation, the method of empirical orthogonal functions, as well as SVD.

### 3.1.2.2 Feature selection

This method is applied in order to reduce the number of features, such as those redundant features from a data set, which can reduce overfitting and consequently improve the accuracy of the predicted model.

FS methods are named under the following titles:

Filter methods—information gain, chi-square test, Fisher score, correlation coefficient, variance threshold.

Wrapper methods—recursive feature elimination, sequential FS algorithms, genetic algorithms.

Embedded methods—L1 (LASSO) regularization, decision tree.

### 3.1.2.3 Featured weighting

FW is technique or a method using a training set that estimates the value of individual features with low and greater influence, that is, features with low influence have a low value, which can be close to zero, while features with greater influence have a greater value.

This method is also sometimes called "Range Scaling & Auto Scaling."

The method increases the data separation in feature space by a simple mathematical transformation, that is, in order to do that, the standard deviation should be known in both $x$ and $y$ for every analyte. The position of every point on the scattergram consequently changes (transforms) to a certain number of classes (Griffin, undated).

The separation can be achieved using the following equation: $(x_1,y_2) \rightarrow (x_1/\sigma_1, y_2/\sigma_2)$ where $\sigma$ is the standard deviation, and $x$ and $y$ are the directions (Box 3.3).

## 3.1.3 Self-organizing map

The self-organizing map (SOM) is an interesting method to be used, however, the final decision about which method should be chosen will be left open for

---

## Box 3.3 Clustering

Clustering is an unsupervised machine learning methodology that aims to partition data into distinct groups, or clusters. There are a few different forms including hierarchical, density, and similarity based. Each have a few different algorithms associated with it as well. One of the hardest parts of any machine learning algorithm is feature engineering, which can especially be difficult with clustering as there is no easy way to figure out what best segments your data into separate but similar groups.

*Sinclair Colin (2018).*

---

the time being until the construction and testing of the device has been completed.

SOMs are data visualization techniques invented by Professor Teuvo Kohonen in the early 1980s (Kohonen, 1981), which reduce the dimensions of data through the use of self-organizing neural networks. The goal here is to discover the underlying structure of the data. Having said that, the kind of structure we are looking for here is very different from that of the PCA (or vector quantization) discussed earlier.

Based on unsupervised learning, the SOM is considered as part of a competitive learning method. That means SOM algorithm does not require any information in addition to the sensor output. SOM is a network formed by $N$ neurons arranged as the nodes of a planar grid.

Kohonen's SOM is called a topology-preserving map because there is a topological structure imposed on the nodes in the network. A topological map is simply a mapping that preserves neighborhood relations (Box 3.4).

---

## Box 3.4 What really happens in SOM?

Each data point in the data set recognizes themselves by competing for representation. SOM mapping steps start from initializing the weight vectors. From there a sample vector is selected randomly and the map of weight vectors is searched to find which weight best represents that sample. Each weight vector has neighboring weights that are close to it. The weight that is chosen is rewarded by being able to become more like that randomly selected sample vector. The neighbors of that weight are also rewarded by being able to become more like the chosen sample vector. This allows the map to grow and form different shapes. Most generally, they form square/rectangular/hexagonal/L shapes in 2D feature space.

---

Unfortunately, in the sensor field, the SOM has not been used as often as other types of data analysis. This is because the interpretation of the result produced by SOM is not straightforward, and requires an accurate reading, despite the fact that SOM is based on an intuitive algorithm. Without the knowledge about the class membership, as it may require in other methods of data analysis, the SOM is able to discover the clustering properties of that set of data. Therefore SOM will be very beneficial for the MOD as it can investigate, as well as understand, various classification properties of multidimensional systems.

### 3.1.3.1 What is SOM and how does it work?

According to Natale, Macagnano, D'Amico, and Davide (1997) the following points provide a brief summary of the SOM:

1. It is a network.
2. This network is made up of $N$ neurons and arranged as the nodes of a planar grid.
3. There are four immediate neighbors for each neuron.
4. The neuron is localized by a vector (e.g., r) and its components are the node coordinates in the grid.
5. Neurons are logic elements—having two states: one output and $m$ inputs, that is, vector z.
6. A real-valued vector is the input.
7. Each output can be active, that is, value 1 or inactive, that is, value 0.
8. An $m$-component code-book vector (e.g., wr) is the characterization of a single neuron, representing the neuron in the input space.
9. Kohonen algorithm (i.e. learning algorithm) modifies the code-book vectors of neurons with every input.
10. The set of code-book vectors (e.g., fwrg) of the grid will be constructed of all z vectors provided.
11. As soon as a new z has been supplied, there are two stages of learning algorithm:
    a. Response: determination of the index (e.g., s) from the condition.
    b. Adaptation: variation of the code-book vectors of all the neurons.
12. The codebook vectors of the SOM neurons will contain the model of the data sample, as soon as the network has attained convergence.

## 3.2 Useful terms

The following terms are relevant to what has been discussed in the previous sections of this chapter. It is useful for the reader to explore them further, including the mathematical formula needed. It is not possible to discuss all the details in this book as they are beyond the scope of the MOD subject. However, some additional materials related to AI have been discussed in Chapter 9, Tests and Training.

### 3.2.1 K-nearest neighbor and K-means clustering

K-nearest neighbor is a simple algorithm that examines all available data and then classifies new set of data that have similarity. What the above means is that by assuming things are in similarity or proximity then KNN algorithm work on it as part of a supervised machine learning. On the other hand, K-means clustering algorithm is unsupervised machine learning, usually works by collecting and grouping into a number of clusters from a set of data, that is, finding similarity.

The distance measures are machine learning algorithms (e.g., KNN, learning vector quantization, SOM, K-means clustering).

### 3.2.2 Distance measures

The distance measures are important aspects in machine learning and some of the commonly used distance measuring are listed below:

- Hamming distance (binary strings/bitstrings): Named after the American mathematician Richard Hamming. The distance between two binary vectors is calculated via the use of this method, that is, the similarity (with the same length of two strings) corresponding characters positions differ is measured. Mainly used for correcting errors or for the valuation of two string or set of data.
- Euclidean distance: Occasionally referred to as "Pythagorean distance." The distance between two real-valued vectors can be calculated using Euclidean distance. While in mathematics, the Euclidean space is measuring the space or line between two points. As a tool, the Euclidean distance provides individual cell's relationship either to certain source or to a set of sources presented via straight-line distance. It is calculated as the square root of the sum of the squared differences between the two vectors.
- Manhattan distance (city block distance): On a 2D plane (or 2D grid), the distance between two points can be measured with a straight line connecting these two points. Some refer to this distance as *the crow flies* distance. Using Pythagoras theorem, we can calculate the distance on 2D plane between the above two points. However, when it is not possible to connect two points via one straight line, then these two points have to follow the grid layout referred to as the *aka taxicab distance*, which is measured between two points along the axes at right angles of the grid. Therefore Manhattan distance is sum of all the real distances (absolute differences) between points across all the dimensions. Manhattan distance is usually preferred over the more common Euclidean distance when there is high dimensionality in the data. It is calculated as the sum of the absolute differences between the two vectors.
- Minkowski distance: Named after the German mathematician Hermann Minkowski. Also known as Minkowski metric and it is the outcome of both Euclidean distance and the Manhattan distance.

- Chebyshev distance: Named after Pafnuty Chebyshev, it is a distance metric, commonly known as the "maximum metric." It measures the distance between two points, that is, two vectors (is the maximum difference/maximum absolute distance) over any of their axis values of two $n$-dimensional points (vectors).

### 3.2.3 Backpropagation

For a brief outlook in this part of the chapter regarding backpropagation, the following provides a brief explanation about this topic.

Backpropagation, which is also referred to as backward propagation, was developed and popularized by David Rumelhart and Ronadl Williams when they proposed it during 1986 to be used for neural network. However, backpropagation was orginally created in the early 1960s and implemented on computers during1970 by Seppo Linnainmaa and Paul Werbos.

Backpropagation is a supervised algorithm that uses delta rule or gradient descent (chain rule) for training multilayer perceptrons in neural network.

Backpropagation is an important mathematical tool, neural network training (tuning of the weights), which is used to improve data prediction accuracy as it performs a backward pass via adjusting the model's parameters, that is, weights, called neuron, adjusted accordingly to obtain accurate outcome in matching predicted/expected result for best model that fits the data. There are two types of backpropagation networks, such as static and reccurrent backpropagation.

A. Backpropagation advantages
  1. One of the fastest and simple method to program.
  2. The inputs are only numbers, that is, no parameters.
  3. No prior knowledge needed related to the network.
  4. It has been accepted as a good standard method.

B. Backpropagation disadvantages
  1. Largely dependent on the data provided.
  2. Error may occur due to noisy data.
  3. Matrix-based approach to be used than mini-batch.

### 3.2.4 Additional terms

Cognitive computing: A method in AI that mimics the human brain function—usually for performing complex operations.

Data curation: Collecting and managing data.

Data mining: The search and discovery of information and patterns in data.

Deep learning: See Chapter 9, Tests and Training.

Ensembling: The work of two or more algorithms, or neural networks, in order to generate accurate outputs, that is, prediction.

Generative adversarial networks: AI algorithm, used within two neural networks working separately for a better outcome.

Machine learning: See Chapter 9, Tests and Training.

Natural language processing: Part of AI for processing and analyzing text and human languages.

Neural networks: See Chapter 9, Tests and Training.

Random forest: Similar to ensembling. This approach is in the form of combining the output of multiple decision trees for better outcome results or prediction.

Reinforcement learning: See Chapter 9, Tests and Training.

Semi-supervised learning: See Chapter 9, Tests and Training.

# References

Brems, M. (2017). *A one-stop shop for principal component analysis*. Medium. <https://towards-datascience.com/a-one-stop-shop-for-principal-component-analysis-5582fb7e0a9c>.

Griffin, M. (undated). *Electronic noses: Multi-sensor arrays*. Davidson, NC: Dept. of Chemistry, Davidson College.

Karakaya, D., Ulucan, O., & Turkan, M. (2020). Electronic nose and its applications: A survey. *International Journal of Automation and Computing*, *17*(2), 179–209. Available from https://doi.org/10.1007/s11633-019-1212-9; <https://link.springer.com/content/pdf/10.1007/s11633-019-1212-9.pdf>.

Kohonen, T. (1981). Automatic formation of topological maps of patterns in a self-organizing system. In I. E. Oja & O. Simula (Eds.), *Proceedings of 2SCIA, scand. conference on image analysis* (pp. 214–220).

Natale, C. D., Macagnano, A., D'Amico, A., & Davide, F. (1997). Electronic-nose modeling and data analysis using a self-organizing map. *Measurement Science & Technology*, *8*, 1236–1243.

Persaud, K. (2016). *Electronic noses and tongues in the food industry—Electronic Noses and Tongues in Food Science* (pp. 1–12). Academic Press. <https://www.sciencedirect.com/science/article/pii/B9780128002438000019#bbib0135>.

Rao, C. R. (1983). Multivariate analysis: Some reminiscences on its origin and development. *The Indian Journal of Statistics, Series B (1960–2002)*, *45*(2), 284–299. <https://www.jstor.org/stable/25052296>.

Rodrigo, D. (undated). *An introduction to multivariate data analysis*. Retrieved from https://towardsdatascience.com/an-introduction-to-multivariate-data-analysis-ece93ceb1ed3. (Accessed 26 August 2020).

Sberveglieri, D. (1999). Metal-oxide semicondictors. In *ASTEQ technologies for sensors*.

Sinclair, C. (2018). *Improving clustering performance using feature weight learning. Towards data science.* Retrieved from https://towardsdatascience.com/improving-clustering-performance-using-feature-weight-learning-d65d4fec77cb. (Accessed 26 August 2020).

Vaefolomeev, S. (1999). Conducting polymer sensors. In *ASTEQ technologies for sensors*.

# Applications

Sensors and machine olfaction devices (MODs) have been developed over the past 40 years (since 1980s) to perform a variety of identification tasks in various governmental, commercial, and industrial fields. However, not long ago, most of the work and publication related to this field were restricted to certain fields or within the area of research. These days, however, various types of commercially available MODs can be purchased anywhere in the world. The reason for the relatively fast development and commercialization of these devices is because they attracted new interest in their application, such as within the fields of food industry, environmental monitoring, medical diagnosis and health monitoring, detection of explosive, manufacturing and laboratory quality control, industrial development and research, detection of illegal drugs, detection of harmful bacteria, space applications, security issues, and other related areas.

The above areas are divided into at least six main branches, and these branches can be subdivided further into many other subbranches or new branches added to it (Fig. 4.1).

The applications of MOD cover a vast area; some of them are still under development and in many cases these applications can overlap each other in various ways. For example, the bacteriological area can be part of the environmental section as well. The same thing can be said about industrial and food applications and likewise between medicines and bacteriological areas (Table 4.1).

The best approach in developing MOD applications for the purpose of obtaining accurate output as well as efficiency, reliability, and longer life cycle is to reduce the number of sensors within the device, that is to narrow the range of applications. However, whether one sensor or two sensors to be allocated or more, the design and the type of materials used in manufacturing a sensor will no doubt provide the level of performance and accuracy needed. A number of sensors presently available in the market may not fulfill all the requirements needed for a higher quality and better performance MOD.

**Introduction to Machine Olfaction Devices.**
DOI: https://doi.org/10.1016/B978-0-12-822420-5.00012-X

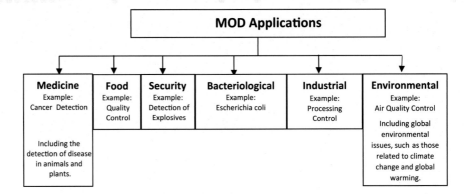

**Figure 4.1**
Areas of possible applications for MOD. *MOD,* Machine olfaction device.

**Table 4.1** Examples of some industry-based applications for electronic noses (Wilson & Baietto, 2009).

| Industry sector | Application area | Specific use types and examples |
|---|---|---|
| Agriculture | Crop protection | Homeland security, safe food supply |
| | Harvest timing and storage | Crop ripeness, preservation treatments |
| | Meat, seafood, and fish products | Freshness, contamination, spoilage |
| | Plant production | Cultivar selection, variety characteristics |
| | Pre- and postharvest diseases | Plant disease diagnoses, pest identification detect nonindigenous pests of food crops |
| Airline transportation | Public safety and welfare Passenger and personnel security | Explosive and flammable materials detection |
| Cosmetics | Personal application products | Perfume and cologne development |
| | Fragrance additives | Product enhancement, consumer appeal |
| Environmental | Air and water quality monitoring | Pollution detection, effluents, toxic spills |
| | Indoor air quality control | Malodor emissions, toxic/hazardous gases |
| | Pollution abatement regulations | Control of point-source pollution releases |
| Food and beverage | Consumer fraud prevention | Ingredient confirmation, content standards |
| | Quality control assessments | Brand recognition, product consistency |
| | Ripeness, food contamination | Marketable condition, spoilage, shelf life |
| | Taste, smell characteristics | Off-flavors, product variety assessments |
| Manufacturing | Processing controls | Product characteristics and consistency |
| | Product uniformity | Aroma and flavor characteristics |
| | Safety, security, work conditions | Fire alarms, toxic gas leak detection |

**Table 4.1** Continued

| Industry sector | Application area | Specific use types and examples |
|---|---|---|
| Medical and clinical | Pathogen identification | Patient treatment selection, prognoses |
| | Pathogen or disease detection | Disease diagnoses, metabolic disorders |
| | Physiological conditions | Nutritional status, organ failures |
| Military | Personnel and population security | Biological and chemical weapons |
| | Civilian and military safety | Explosive materials detection |
| Pharmaceutical | Contamination, product purity | Quality control of drug purity |
| | Variations in product mixtures | Formulation consistency and uniformity |
| Regulatory | Consumer protection | Product safety, hazardous characteristics |
| | Environmental protection | Air, water, and soil contamination tests |
| Scientific research | Botany, ecological studies | Chemotaxonomy, ecosystem functions |
| | Engineering, material properties | Machine design, chemical processes |
| | Microbiology, pathology | Microbe and metabolite identifications |

# 4.1 Medicine

The use of the MOD device has several major advantages over conventional diagnostic tools and has great potential as a method to detect diseases, and other types of illnesses. The metabolic processes occurring in our bodies generate volatile organic compounds (VOC) in the form of gases within various cells. This may take place during abnormal functions, such as infection or other types of illnesses. A clinical diagnostic, therefore, such as detection of glucose, cholesterol, cancer, and a variety types of noninfections or infections illnesses and diseases can be detected via the use of MOD.

One of the earliest examples of the usage of MOD in the medical field was by C. William Hanson, III, MD, at the University of Pennsylvania Medical Center, which focused on the use of the e-Nose to find out about the presence of pulmonary infection. He took breath samples from the ventilators of 19 patients by collecting them in plastic bags; about half of them (9) were suffering from pneumonia. These breath samples were then fed into the *aroma-analysis device* (i.e., e-Nose). Each sample was analyzed and the result (pattern) for each sample displayed on the screen. This electrical resistance distinguished infected patients from noninfected patients using different patterns (Box 4.1).

When an MOD device is capable of displaying accurate results immediately, a device such as this can help in reducing our dependency on various diagnostic machines such as "X-rays" and waiting for the result of medical sample analysis

## Box 4.1 Examples of e-Nose medical applications

Recent advances in clinical e-nose applications have also been achieved in the detection of patients' drug use, exposure to hazardous gases and toxins, organ failure, and physiological abnormalities. Metal-oxide semiconductor-type e-nose sensors have been used for many of these applications for detection of abnormal gases released from the lungs (in breath), mouth cavity, skin, and headspace VOCs from various body fluids. Specialized atypical e-nose sensor types such as field-effect transistors, microelectromechanical systems, ion-selective electrodes, silica-gel nanospheres, and electrochemicals were utilized for these clinical
gas-detection diagnostic applications mostly from the analysis of VOCs and toxic metals present in body fluids.

*Wilson (2016)*

to arrive from the lab before diagnosis can be made by the doctor; in addition it will reduce the use of antibiotic in many cases.

An odor from the body can be identified using an MOD; this can be from a wound, breath, or body fluid. Therefore gastrointestinal and sinus infections, diabetes, liver problems (odors coming from body fluids can indicate liver and bladder problems) and many more ailments can be detected by simply analyzing the odors in the breath or from other parts of the body.

From the year 1998 there were already a number of noninvasive eNoses, which dealt with various medical conditions in the form of detecting and analyzing—such as *Electronic nose analysis of urine samples containing blood* (Di Natale et al., 1999); *Bacteria classification using Cyranose 320 electronic nose* (Dutta, Hines, Gardner, & Boilot, 2002); *The prediction of bacteria type and culture growth phase by an electronic nose with a multi-layer perceptron network* (Gardnery, Craveny, Dowz, & Hinesy, 1998); *Human skin odor analysis by means of an electronic nose* (Di Natale et al., 2000). The above publications, of which most of them are more than 20 years old, are just few examples of some of the research that has been done in the past, all relating in one way or another to the medical or biological field using MOD. Presently this kind of research within the above field has been multiplied many folds as advances in MOD have made it possible to expand this work further.

Other examples of areas where MOD can be used in the medical field effectively and in a noninvasive manner can be cancer, diabetes, various types of infections, arthritis, and many similar types of illnesses and diseases. The development took place over the past 20 years on the MOD device, which managed to improve its performance in various ways compared to the older devices manufactured at the end of the last century. This is mainly due to the development of the sensor and the new generation of software programs used in the device.

It has been accepted by a number of researchers and developers in this field that the present and future function for the MOD in medicine is the tele-surgery.

There are many advantages in using the MOD device in the medical field. These advantages can be summarized in the following points:

1. It can provide instant result(s).
2. Consequently, can save time and costs.
3. Reducing the reliance on harmful machine (e.g., X-rays) during diagnosis.
4. Mobility (in the case of a mobile MOD).

There are no real disadvantages in using MODs in the medical field apart from the "further reliability," which has not been established yet, simply because these devices are continuously undergoing research and development at the present time.

The most advantageous area in which MODs could be used is the examination and diagnosis of lung problems. For this reason, it was one of the first applications considered for the device at the very early stages of development (Table 4.2).

Medicine, therefore, has recognized that some illnesses/infections produce unpleasant characteristic odors, for example diabetes and some hepatic diseases. Hence MOD will be an ideal device that can use these phenomena in order to recognize various types of medical conditions at a much faster pace than what is taking place at the present time. Lung and stomach diseases (plus similar types of diseases and illnesses) can be analyzed by taking a sample (or samples) from the patient's breath and exposing/injecting them into MOD for the purpose of analyzing the type and intensity of the illness/disease (Table 4.3).

## 4.2 Food

At present, the food industry is one of the biggest potential markets for MODs. This is already hap at various companies across the globe. The replacement of panels of human experts, using MOD, can work to the advantages of the food/drink companies in various ways, such as accuracy, time, and cost. MOD, in some cases (in particular when qualitative result will do), can be used to reduce the amount of analytical chemistry performed during food production (Box 4.2).

A few good examples of the applications of MOD in this vast area of use are "inspection of food, for example, process control," "food grading quality, for example, freshness and spoilage," "fermentation control," "fish and meat inspection, for

**Table 4.2** Applications of MOD devices in clinical practice (Wilson, 2016).

| Applications | Organ system | Detection | E-nose model system | Sensor type and number |
|---|---|---|---|---|
| Disease diagnosis | Blood | Pathogen (*Salmonella*) | DR-ADF $SiO_2$–$TiO_2$ biosensor | MOS-BS |
| | Blood | Malaria | Experimental CNF on NMBs | CNF-BS |
| | Breast | Cancer | BreathLink (GC + SAw) | SAw 1 |
| | Colon | Cancer | Cyranose 320 | CP 32 |
| | Foot | Microbial infections | Cyranose 320 | CP 32 |
| | Lung | Cancer | Many types: gold nanoparticles, CP, MOS, SAw | Various |
| | Lung | Asthma, CF, COPD, TB | Bloodhound 114 Cyranose 320 MOSeS ii e-nose | CP 14 CP 32 MOS 8, QMB 8 |
| | Ovaries | Cancer | $SnO_2$, $TiO_2$ Figaro TGS2600-series | MOS 16 |
| | Prostate | Cancer | ChemPro 100i | iMS 1 |
| | Upper airway | OSAS | Cyranose 320 | CP 32 |
| | Urinary | Pathogens | iMS Cell-e-nose | iMS 8, MOS 6 |
| Drug use | Mouth | Alcohol (ethanol) | experimental $SnO_2$/ZnO core-shell nanorods | MOS |
| | Skin | Marijuana (*Cannabis*) | Experimental $SnO_2$ | MOS 3 |
| Hazardous gas exposure | Lungs | CO, $NO_2$, $NH_3$, $SO_2$ | $in_2O_3$[a], $SnO_2$[a] CNT-based FeTs | FeT |
| Organ failure | Kidney | Ammonia breath analysis | TGS 2602 | MOS 1 |
| Patient iD | Body | Human odors | Experimental TGS (Tagushi) e-nose | MOS 5 |
| Physiological | Body fluids | Health parameters | CardioMeMS, others | MeMs |
| | | Antioxidant capacity | TGS 2602 eC biosensor | eCB |
| | | Urea buildup | Silica-gel $SiO_2$NPs biosensor | SG-BS |
| Toxin exposure | Body fluids | Mercury poisoning | Experimental iSe | eC |

*Source*: Adapted and edited from Wilson, A. (2016). Recent progress in the design and clinical development of electronic-nose technologies. Review. *Nanobiosensors in Disease Diagnosis*. Open Access. https://www.dovepress.com/front_end/cr_data/cache/pdf/download_1614694253_603e476d26e44/NDD-66278-recent-progress-in-the-design-and-clinical-development-of-el_012216.pdf.

**Table 4.3** Biomarkers that MOD can detect and sources of possible type of diseases (Thalakkotur, Prabhahari, Sukhananazerin, & Biji, 2015).

| Biomarkers | Sources | Diseases |
|---|---|---|
| Ammonia | Protein metabolism | Kidney failure, cirrhosis or hepatitis, hepatic encephalopathy, peptic ulcers, halitosis and asthma |
| Acetone | Acetoacetate decarboxylation | Diabetes, lung cancer, dietary fat losses, congestive heart failure, brain seizure |
| Isoprene | Mevalonate metabolism | Disorders in cholesterol metabolism such as hypercholesterolemia |
| Methane | Intestinal bacteria metabolism of carbohydrates | Intestinal problems, colonic fermentation |
| Hydrogen sulfide | Metabolism of L-cysteine, cystathionine beta-synthase (CBS) | Airway inflammation |
| Nitric oxide | Nitric oxide synthase | Asthma, acute lung injury, ARDS, inflammatory lung diseases, lung infection, lung cancer, rhinitis |
| Ethane | Peroxidation of polyunsaturated fatty acids | Oxidative stress, vitamin E deficiency, breast cancer, active ulcerative colitis |
| Pentane | Peroxidation of polyunsaturated fatty acids | Liver diseases, schizophrenia, breast cancer, rheumatoid arthritis, oxidative stress, acute myocardial infarction, asthma |

*Source*: From Thalakkotur, M., Prabhahari, P., Sukhananazerin, A., & Biji, P. (2015). Technologies for clinical diagnosis using expired human breath analysis. *Diagnostics*, *5*, 27â€"60. Open Access. https://doi.org/10.3390/diagnostics5010027; https://www.mdpi.com/2075-4418/5/1/27/htm.

---

## Box 4.2 Off-flavors in foods

Off-flavors in foods originate mostly from bacterial and/or fungal metabolism and several studies have been carried out on the detection of the volatile chemicals produced from the microorganisms mentioned. The isolated species include gram-negative bacteria, gram-positive bacteria, and several fungi. The risk of fungal contamination is also related to several mycotoxigenic species that produce mycotoxins very harmful to human and animal health.

*Casalinuovo, Pierro, Coletta, and Francesco (2006)*

---

example, lipid oxidation," "beverage container inspection, for example, packaging," "flavor control, for example, taints and off-flavor," and many more (Box 4.3). The MODs therefore can bring about a revolution within the food industry in general and food quality control, in particular.

---

### Box 4.3 Fruit aroma

Differences in the aroma characteristics of different fruit varieties are attributed to variations in their chemical profiles, based on the types of VOCs present and the relative concentrations of individual volatiles found in the aroma mixture. The principal VOCs, found in the aromas of specific fruit cultivars, may be used to distinguish between different fruit varieties. These differences in aroma composition and relative abundance of fruit volatiles in different fruit varieties are the means by which e-nose devices are capable of recognizing differences in fruit aromas and discriminating between fruit cultivars based on their distinct aroma signature patterns resulting from variations in e-nose response (sensor-array outputs) to different fruit aromas. Thus e-nose discriminations of fruit aroma are determined both by the volatiles present and molar ratios of individual components found in each aroma (gaseous mixture/s).

*Baietto and Wilson (2015)*

---

The food industry has many specialized areas where large number of research projects and applications are presently taking place, at various research institutes and within various international companies. The sampling procedures are a problem when it comes to food analysis controlled by MOD system, as the device measures an image rather than measuring the sample in question. The above can be solved when using deep learning, as a way in training the device (see Chapter Tests and training).

The competition among industries and researchers for specialized MODs in this field is very high. For this reason, the possibility of doing similar or even the same kind of research is very high as well. Therefore it would be advisable to avoid choosing an application in this area purely from academic research approach, unless combine both business and academic applications at the same time.

Finally, MOD is vital in the detection of: ochratoxin (Aspergillus and Penicillium fungi), aflatoxin (Aspergillus mold), mycotoxin (toxic chemical produced by Fusarium, Aspergillus, and Penicillium), *Salmonella* (*Salmonella enteritidis* and *S. typhimurium*), *Escherichia coli* (intestinal bacterium).

## 4.3 Security

Security, or defense, as it is sometimes referred to, is another area where interest in the MODs is being researched presently by defense and securities institutes and organizations.

Present methods are used to check for explosives, for example, at airports and other checking security points; personal and devices could be replaced with mobile MODs. This means the use of X-rays, metal detectors, security guards, and trained dogs, under certain conditions, can no longer be as accurate and fast in the detection process, as those provided by mobile devices.

According to the authorities at the airports in the London area, the most difficult explosive to detect is a "plastic" explosive, as this type of explosive is non-metallic, so it does not show up in X-ray machines (Sample, 2010) (Box 4.4).

For this reason, MOD will be the ideal way in detecting these kind of materials—consequently greatly aiding security at all airports and similarly sensitive

---

## Box 4.4 Detection of explosives

Detection of explosives and nerve agents has been achieved through several approaches including doped or surface-modified sensitive materials, regulated contact barrier, absorbent materials with unique pore size. Hierarchical ZnS nanostructure doped with $Mn^{2+}$ at various concentration can be a successful tool when they assemble together as a sensor array to discriminate explosives.

*Hu et al. (2018)*

---

locations.

A security check-up, via the use of MOD, can be used on cargo. Using the above mobile devices, a security check can be easily made before take-off and/or before landing, as the checking can take place in real-time. This method will save lot of time, drastically lower present security costs, and may save lives.

MOD can play an important and vital role in preventing major possible terrorists attack or, even can minimize the impact.

Therefore MOD can be successfully used on "screening of cargo containers," "monitoring ambient air in the government buildings," "monitoring ambient air in subways," "profiling ambient air in commercial airlines," and checking various types of explosives.

Finally, MOD can detect chemical warfare, such as nerve gases, carbamate pesticides, organophosphorus insecticides, as well as hydrogen cyanide.

## 4.4 Bacteriological

MODs have been used to detect volatiles emitted by growing bacteria (including viruses) using concentrated samples. These devices have consistently been able to distinguish between various arbitrarily samples of bacteria.

In a research study titled *The prediction of bacteria type and culture growth phase by an electronic nose with a multi-layer perceptron network* (Gardner et al., 1998), the bacterial culture head space has been used after bacteria was grown in a standard nutrient medium. According to Gardner et al. (1998), using an array of six different metal oxide gas sensors it was found to successfully classify 100% of the unknown *S. aureus* samples and 92% of the unknown *E. coli* samples.

The above illustrates that the MOD can be used successfully to detect varieties of microbiological organisms. The detection can be used, therefore, in a similar and reversed way, where certain components are needed to grow certain types of bacteria under certain artificial conditions (please refer to environmental applications below) (Box 4.5).

The bacteriological applications of MOD therefore may vary, and certainly can generate useful results in this field. However, as mentioned earlier on, the device

---

## Box 4.5 Bacteria and VOC

Many strains of bacteria can produce the same VOCs but with different concentrations. On the other hand, some VOCs can exist in the specific bacteria metabolism pathways. For example, the concentration of ethanol in gram-positive bacteria is more than gram-negative bacteria, which is inverse for acetaldehyde and formaldehyde. Propanol, pentanol isomers, acetic acid, and dimethyl disulfide were shown only in the *E. Coli* VOCs profile while butyl butyrate, 3-Methyl butyrate and 2-Methyl butyrate were indicated in chemical composition of *Staphylococcus* strains.

*Bordbar, Tashkhouriana, Tavassoli, Bahramali, and Hemmateenejada (2020)*

---

applications can overlap other areas; therefore these applications can be part of the medical field as well, depending on the purpose sought after.

## 4.5 Industrial

Just as in other fields, the MODs within various types of industries are already playing a major role. In fact, the MOD or the electronic nose idea started from here, in particular from the space industry where the initiative of producing a device with sensors for gases (for the detection of toxic gases) formed the basic principle of electronic nose later on.

In 1982, when Persaud and Dodd introduced the actual concept of an electronic nose, a device was designed comprising an array of essentially nonselective sensors and an appropriate pattern recognition system, marking the first device labeled as electronic nose. Nowadays the concept of the MOD is being applied in various sectors of the industry across the globe, covering a wide range of products and services, that is, before, during, and after the process of industrial/product applications, to mention few, such as the detection of industrial toxic gases, hydrazine fuels, chemical warfare gases (e.g., stored mustard gas). (Box 4.6).

Various industries in different parts of the world do manage the emissions of various obnoxious emissions and gases via treatment plants, as part of their legal obligations and efficiency as well as part of their sustainability

---

## Box 4.6 Industrial emissions and VOCs

One of the major problems in industrial emissions analysis is the detection of mixture of VOCs. It is inevitable that industrial pollutants will have only single VOC emitted. This difficulty arises from primarily two factors: first, in many cases the sensor may have a very similar response to two very different compounds, a direct result of the inadequate specificity of the sensor. Second, the sensor response to one of the components of the mixture may be so high that the response to the other components may be completely masked. Besides the above snags, an electronic nose for industrial applications needs to deal with fast recovery and the measurement accessibility in remote areas. High sensitivity is necessary for effective analysis and the subsequent development of treatment technologies.

*Deshmukh, Bandyopadhyay, Bhattacharyya, Pandeya, and Jana (2015)*

---

approach. MOD began to replace some of the old approaches and to monitor various sections of these industries as part of managing and controlling the above issues.

Industrial emission in the form of gases such as acetaldehyde ($CH_3CHO$), acetylene ($C_2H_2$), antimony pentachloride ($SbCL_5$), arsine ($AsH_3$), carbon monoxide (CO), ethanol ($C_2H_5OH$), nitrogen dioxide ($NO_2$), sulfur dioxide ($SO_2$) can be detected even at low concentration, specifically close to or within an industrial and commercial site manufacturing or processing substances that may have impacts on human health and the natural environment.

## 4.6 Environmental

Large amounts of wastes, in particular dangerous waste and by-products waste, such as nuclear, chemical, and biological, have been generated over a long period of time by heavy industrial production all over the globe, which should have been dealt with safely and immediately. However, in many places around the world this is not the case. Many of these wastes are toxic, and in particular waste generated from weapon production. The applications of MOD can help immensely in dealing with waste in general and minimizing or preventing negative impacts on the environment. MOD can be also used to measure/monitor air quality, water quality, process control (prevention tools), and pollution/odor control.

The role of a mobile MOD is vital when it comes to environmental issues and sustainability. This is because MOD can identify contaminants in real-time, long before having an adverse, irreversible impact on our environments.

This is one of the reasons that the development of sensor technology has widened the interest in the use of characteristic volatile and odorous compounds produced by various organic and nonorganic materials in nature. Obviously, many of these MODs are used for an early indication of the deterioration in various materials, such as the oil used as insulator for the electrical transformers. Sensor arrays to measure traces of volatile chemicals could detect early fungal growth in libraries and archives. MODs could also offer a rapid and relatively simple technique for continuous monitoring of water quality. Therefore taste and odor, which is caused by chemicals dissolved in the water supplied to us, can be detected and identified using a variety of analytical techniques and sensory methods. MODs, such as sensor arrays, have been used to monitor the environment in areas such as wastewater and nuisance odor assessment (Box 4.7).

---

### Box 4.7 Water quality monitoring.

One of the most important legislative limits fixed for wastewater treatment plants is the so-called "Biological Oxygen Demand" (BOD) concentration. The BOD is a parameter that describes the degree of water pollution, and it represents the difference between the quantity of dissolved oxygen in a water sample before and after an incubation period of 5 days, at 20°C, in presence of a bacteria flora. The analysis of the BOD therefore indicates the contents of organic biodegradable matter in wastewaters, expressed in terms of quantity of oxygen needed for the aerobic biodegradation by microorganisms. For this reason, the first studies regarding the application of electronic noses for water quality monitoring focused on the measurement of this parameter.

*Capelli, Sironi, and Del Rosso (2014)*

---

Various research studies in this area have demonstrated the potential application of using MODs as a means for a continuous early warning monitoring system, such as in the above example, that is, for taste and odor. These applications are still not used on a large scale; this could be due to a number of factors, including some of the applications outcome of the MOD system in relation to environmental areas, that is, the limitation of certian parts of the device hardware when exposed to open environment, which means further development and research are still needed before the above device can be used in an efficient and reliable way.

When it comes to the environment, MODs can be applied in areas such as agriculture, animals and animal's environment, water and sewage systems, atmospheric conditions, mines, caves, and many other areas.

The MOD design in this book can choose from a number of areas for the final applications of the device. These areas were subdivided further until a very specialized environmental section is selected. This can be done directly via the use of the device by choosing the required application, if the device is programmed for more than one application.

For example, there are various types of environmental chemical pollution that the MOD can be set-up to detect. Depending on the application chosen within the device, the following are some examples (Wilson, 2012):

- Inorganic chemicals (e.g., oxides of carbon, nitrogen, sulfur, mercury).
- VOCs (e.g., carboxylic acids, organic acids, alcohols, aldehydes, and carboxylic acids).
- Other sourced chemicals (e.g., toxic industrial chemical, agricultural pesticides, chemical warfare agent, waste-treatment, composting).
- Other toxic compounds (e.g., benzene, toluene, xylene, sulfur compounds).

Looking at natural and artificial atmospheres, the idea was narrowed down to the following sections in which MOD can be used to replace older inefficient and/or costly methods used at the present time. These areas are:

1. Commercial and noncommercial airplanes.
2. Submarines.
3. Spacecraft/space shuttles.
4. Chambers/incubators (for microorganisms).

The MOD can be used to check up the existent components, which make up the required/standard elements needed to create and maintain certain artificial atmospheric conditions.

These artificial atmospheres were produced to achieve known conditions for the purpose of obtaining certain result(s). The artificial environments were usually created in different areas for biological and nonbiological purposes, as well as to provide conditions for human habitat.

The above can be narrowed down further to point 4 (chambers/incubators for microorganisms) as a more likely area for the MOD applications, which this book is providing.

Finally, for the space program, such as the environment within the international space station as well as the shuttle, MOD is one of the vital devices for the safety and the survival and monitoring the well-being of the crew. The following can be applied as well to the above four areas mentioned above, partly or fully:

1. Air contaminants.
2. Airlocks hypergolic propellant contaminants.
3. Impending fire notification.
4. Stored food quality and health safety.

# References

Baietto, M., & Wislon, A. (2015). *Electronic-nose applications for fruit identification, Ripeness and Quality Grading*. Sensors (Basel). Available from https://doi.org/10.3390/s150100899. <https://www.ncbi.nlm.nih.gov/pmc/articles/PMC4327056>, Accessed 06.11.2015.

Bordbar, M., Tashkhouriana, J., Tavassoli, A., Bahramali, E., & Hemmateenejada, B. (2020). *Ultrafast detection of infectious bacteria using optoelectronic nose based on metallic nanoparticles*. Elsevier. <https://www.sciencedirect.com/science/article/pii/S0925400520306079>, Accessed 05.09.20.

Capelli, L., Sironi, S., & Del Rosso, R. (2014). Electronic noses for environmental monitoring applications. *Sensors, 14*, 19979–20007. Available from https://doi.org/10.3390/s141119979. <https://core.ac.uk/reader/55254150>, Accessed 05.09.20.

Casalinuovo, I., Pierro, D., Coletta, M., & Francesco, P. (2006). Application of electronic noses for disease diagnosis and food spoilage detection. *Sensors (Basel), 6*(11).

Deshmukh, S., Bandyopadhyay, R., Bhattacharyya, N., Pandeya, R., Jana, A. *Application of electronic nose for industrial odors and gaseous emissions measurement and monitoring – An overview*. (2015). <https://www.sciencedirect.com/science/article/pii/S0039914015300904> Accessed 05.09.20.

Di Natale, C., Macagnano, A., Paolesse, R., Tarizzo, E., Mantini, A., & D'Amico, A. (2000). Human skin odor analysis by means of an electronic nose. *Sensors and Actuators B, 65*, 216–219.

Di Natale, C., Mantini, A., Macagnano, A., Antuzzi, D., Paolesse, R., & D'Amico, A. (1999). Electronic nose analysis of urine samples containing blood. *Physiological Measurement, 20*, 377–384.

Dutta, R., Hines, E. L., Gardner, J. W., & Boilot, P. (2002). Bacteria classification using Cyranose 320 electronic nose. *Biomedical Engineering Online, 1*(4). Available from https://doi.org/10.1186/1475-925X-1-4.

Gardner, R. (1998). The prediction of bacteria type and culture growth phase by an electronic nose with a multi-layer perceptron network. *Measurement Science & Technology, 9*(12). Available from https://iopscience.iop.org/article/10.1088/0957-0233/9/1/016.

Gardnery, J. W., Craveny, M., Dowz, C., & Hinesy, E. L. (1998). The prediction of bacteria type and culture growth phase by an electronic nose with a multi-layer perceptron network. *Measurement Science & Technology, 9*, 120–127.

Hu, W., Wan, L., Jian, Y., Ren, C., Jin, K., Su, X., … Wu, W. (2018). Electronic noses: From advanced materials to sensors aided with data processing. *Advanced Materials Technologies*. Wiley online library. <https://www.onlinelibrary.wiley.com/doi/full/10.1002/admt.201800488>, Accessed 04.09.20.

Persaud, K., & Dodd, G. (1982). *Nature, 299*, 352−355.

Sample, I. (2010). Why PETN explosive is hard to detect. Guardian Newspapers. Cited from "The Hindu" website: < https://www.thehindu.com/sci-tech/Why-PETN-explosive-is-hard-to-detect/article15675553.ece > .Accessed 04.09.10

Thalakkotur, M., Prabhahari, P., Sukhananazerin, A., & Biji, P. (2015). Technologies for clinical diagnosis using expired human breath analysis. *Diagnostics, 5*, 27−60. Open Access, Available from https://doi.org/10.3390/diagnostics5010027, <https://www.mdpi.com/2075-4418/5/1/27/htm>.

Wilson, A. (2012). Review of electronic-nose technologies and algorithms to detect. *Procedia Technology, 1*(2012), 453−463. Open access under CC BY-NC-ND license. <http://www.sciencedirect.com>.

Wilson, A. (2016). Recent progress in the design and clinical development of electronic-nose technologies. Review. *Nanobiosensors in Disease Diagnosis*. Open Access. <https://www.dovepress.com/front_end/cr_data/cache/pdf/download_1614694253_603e476d26e44/NDD-66278-recent-progress-in-the-design-and-clinical-development-of-el_012216.pdf>, https://www.srs.fs.usda.gov/pubs/ja/2016/ja_2016_wilson_001.pdf.

Wilson, A. D., & Baietto, M. (2009). Applications and advances in electronic-nose technologies, Review*Sensors, 9*, 5099−5148. Available from https://doi.org/10.3390/s90705099.

# Biosensors and machine olfaction device design

This chapter provides an outline for a new mobile MOD design, characterized by a longer working life cycle and for continuous sensor functioning, including the time needed for sensor replacement, in addition to the discussion related to biosensors. This also means that the device can be programmed to be able to monitor on short and long timescale, depending on the chosen application.

Available MOD devices may either lack mobility or they are not designed specifically for different applications. Most of these available devices use "arrays of sensors," that is, eNoses, rather than a device with two active sensors, as in the case with the new design provided in this chapter. In addition, many of the eNoses available in the market do not show the concentration of one type of compound in a sample; however, combining two types of active sensors (biosensor, e.g., proteins of olfactory receptors and technical sensors) can solve this issue. It means a combination of bioelectronic nose with the design of MOD in one device. There are also additional sensors which can be activated, partially or fully, whenever the need arises.

## 5.1 Biosensors

Chapter 1, Background, materials, and process, and Chapter 2, Comparison and validations, dealt with technical/chemical sensors in general. In this chapter, biosensors are presented to be incorporated with the new design as they are highly sensitive compared with other types of sensors as well as highly selective regarding the detection of specific analyte in a sample.

Broadly speaking the difference between a biosensor and a technical/chemical sensor is simply the detection part is sourced from biological materials. The two main parts of a biosensor are the bioreceptor and transducer.

According to Malhotra and Pandey (2017), for the fabrication of a biosensor for nonspecialist markets, certain conditions are required, as follows:

- The desired analyte should be specific and stable under a normal storage condition.

**Introduction to Machine Olfaction Devices.**
**DOI:** https://doi.org/10.1016/B978-0-12-822420-5.00004-0

- The sensor should be accurate and precise and show high sensitivity in a reproducible way, and linearity must be obtained with different concentrations.
- Physical parameters such as pH and temperature should be optimized, which will lead to sample analysis with minimal pretreatment.
- The biosensor should be small and biocompatible so that it can be used for invasive monitoring in clinical diagnostics.
- The fabricated biosensor should be portable, cost-effective, small, and capable of being used by semiskilled operators.
- The biosensor should provide real-time analysis.

Biosensors broadly mimic certain aspects of the human sense of smell, where sensitivity, selectivity, stability, reproducibility, and linearity are all important part of a reliable sensing mechanism. More importantly, the low cost of the biosensors compared with other type of sensors is another advantage that should be considered, as cost is a major factor when it comes to general and wider MOD use.

There are many fields/applications where biosensors can be used effectively, such as food quality monitoring, water quality management, environmental monitoring, toxins of defense interest, drug discovery, prosthetic devices, and disease detection.

In general, the mode of operation of biosensors is categorized as catalytic or affinity. The catalytic mode employs an active biocomponent or active polynucleotides, which can react with an analyte, while the affinity biosensors depend on the analyte interaction (molecule) with a biological receptor (Pethig & Smith, 2013).

How does a biosensor work? An analyte interact with a receptor (e.g., a cell, aptamers, DNA, antibodies, and enzymes). Depending on the concentration of the analyte used the interaction or the outcome can be in the form of light, heat, pH, charge, or mass change. The transducer (e.g., quartz electrode) will convert the aforementioned forms of energy into electrical (or optical) signals. The signals will be amplified and processed (analog to digital) in a computer or electronic device and then displayed on a monitor.

## 5.1.1 Brief history

More than six decades ago, that is, during 1956, an American biochemist, Professor Leland Clark (Professor of Chemistry) published his article on the oxygen electrode, which was the approach for the first biosensor, designed for the purpose of measuring oxygen in the blood, such as during cardiac surgery. Further development of the biosensor by Clark and Lyons took place during 1962 (glucose biosensor).

During 1969 Guilbault and Montalvo constructed the first potentiometric biosensor to detect urea. Divies during 1975 proposed the usage of bacteria as the biological element for microbial electrodes to measure alcohol. Further development of the biosensor continued to take place during the past decades, attracting large investment in this field (Table 5.1).

Today' usage of biosensors covers a wide range of applications. Present and future advancement in nanotechnology and its implementation in the production process of biosensors already (and will) provides much better outcome in the form of applications and accurate results as well as in reducing cost. Currently, there are a variety of biosensors, and as a result, accuracy, application and cost differ accordingly (Fig. 5.1).

**Table 5.1** Biosensors development important dates.

| | |
|---|---|
| 1970 | Discovery of ion-sensitive field-effect transistor (ISFET) by Bergveld |
| 1975 | Fiber-optic biosensor for carbon dioxide and oxygen detection by Lubbers and Opitz |
| 1975 | First commercial biosensor for glucose detection by YSI |
| 1975 | First microbe-based immunosensor by Suzuki et al. |
| 1982 | Fiber-optic biosensor for glucose detection by Schultz |
| 1983 | Surface plasmon resonance (SPR) immunosensor by Liedberg et al. |
| 1984 | First mediated amperometric biosensor: ferrocene used with glucose oxidase for glucose detection |
| 1990 | SPR-based biosensor by Pharmacia Biacore |
| 1992 | Handheld blood biosensor by i-STAT |

*Source*: Adapted and edited from Bhalla N., Jolly P., Formisano N., & Estrela P. (2016). Introduction to biosensors. *Essays in Biochemistry 60*, 1–8, DOI: https://doi.org/10.1042/EBC20150001 (Bhalla, Jolly, Formisano, & Estrela, 2016).

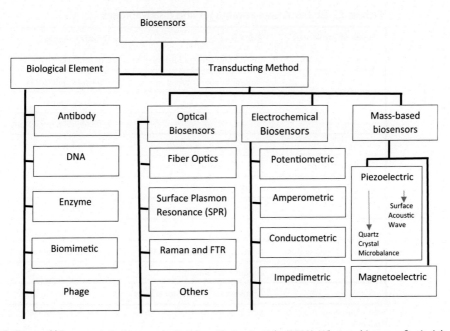

**Figure 5.1** Types of biosensors. *Redrawn and edited from Electronics Hub. (2019).* What are biosensors? principle, working, types and applications. *Retrieved from https://www.electronicshub.org/types-of-biosensors/. (Accessed 27 October 2020) (Electronics Hub, 2019).*

The source of the original materials for these biosensors may have different advantages and disadvantages, including lifetime, in general; however, the main advantage for most biosensors is sensitivity to odor (Table 5.2).

### 5.1.1.1  Biological cell

A question may arise on why there is the need to use biosensor and, as such, why a biological whole cell can be a good option? First, a good knowledge of cells' function is essential. This is mainly due to the complexity involved at this level. Despite these difficulties and the shorter life cycle of a cell, functional and analytical information are vital, which can be obtained directly from functional cells. These kinds of information are important within the medical and pharmaceutical fields, as well as in relation to environmental issues. The usage of biological cells, such as bacterial components or genetically modified bacteria or cells with certain type of receptors, can function as sensors, such as for the investigation of environmental pollution. These biological sensors can outperform other types of sensors in their sensitivity and accuracy as a part of an MOD system. Having said that, the manufacturing process entails the combination of these biomaterials on processor chip, that is, using microprocessing fabrication or, even better, nanoscale fabrication. If the latter is used, then nanosensors performance exceed other types of sensors, in addition to the lower costs entailed in manufacturing them. Low-cost biosensors are essential as these will be disposed due to the lifecycle of the organic matter used.

**Table 5.2**  Biosensors advantages and disadvantages.

| Types of biosensor | Advantages | Disadvantages |
|---|---|---|
| Tissue-based sensor | Ease to fabricate and immobilize | Lack of specificity |
| | Natural odorant profiles | Individual differences |
| | Low cost | Olfactory fatigue |
| | | Difficulty of natural state and storage |
| | | Need to kill animals |
| Cell-based sensor | High sensitivity and selectivity | Difficulty of handling |
| | Single type of ORs | Necessity of culturing |
| | Nature of membrane for ORs | Low stability |
| | | Difficulty of long-term sensing |
| Receptor-based sensor | High sensitivity and selectivity | Difficulty of purification and isolation |
| | Longer-term stability | |
| | Acquirement of receptor activity | |

Source: Nakamoto, T. (2016). *Chapter 2: Essential of machine olfaction and test* (p. 47). Wiley, ISBN: 9781118768488 [Nakamoto (2016)].
OR, olfactory receptors.

Biological (as well as technical/chemical) sensors for various kinds of devices, including MODs, should be manufactured using nanotechnology approach, that is, on a large commercial scale for the production of nanosensors. As mentioned previously, this is because the benefits exceed in various ways the present sensors available on the market. According to Aliofkhazraei (2014), most common applications of nanosensors are summarized by the following points:

- detection of chemical and biological agents and quantities;
- DNA sequencing;
- diagnosis of illnesses and drug synthesis;
- effective quick tests on new drugs;
- portable control systems to maintain health of agricultural and food products in stocks and transportation processes;
- integrated nanosensor systems of intelligent measuring, reporting, and control of plants or livestock;
- high-precision biosensoring for detection of proteins; and
- quick diagnosis of pathogenic agents.

Examples of nanosensors are metal nanoparticle–based, nanostructured metal oxide–based, carbon nanotube–based, graphene–based, and quantum dot–based nanosensors.

As compact devices, biosensors contain biological substance connected to a transducer. The transducer type may vary, depending on the type of system design and possible applications; therefore it can be an optical, electrochemical, piezoelectric, or thermometric device. Optic biosensors are a group of electromagnetic tracers with an optic output signal. In mechanical biosensors, usually, there is a biosensitive element attached, that is, an antibody sonic wave transducer, and functions via the change in resonant frequency. Electrochemical biosensors is made from metallic nanoparticles, that is, vast number of biomolecules spattered by them without interfering with their biological functions.

Cellular biosensors can be a single biological cell, such as a natural bacterium or in the form of genetically modified cell. Cellular sensors can be used for either within the medical field or environmental pollution control.

As in technical/chemical sensors, biosensors parameters should be considered carefully, such as factors related to noise, selectivity, sensitivity, transfer function, format, drift, precision and accuracy, detection limit and decision limit, dynamic range, response time, resolution, bandwidth, hysteresis, effects of pH, effect of temperature, and testing of antiinterference (Pethig & Smith, 2013).

## 5.1.2 The device

A biosensor or biological sensor is a device that employs a living organism or certain biological molecules, such as protein or enzymes for the purpose of detecting the presence of chemicals. It is made up of biological (e.g., antibody, enzyme, and nucleic acid) and physical (e.g., transducer, amplifier, and processor) components. Therefore as an analytical device, a biosensor can detect certain changes taking place within a biological process and, as a result, generates an electrical signal. Biosensors are listed as electrochemical, optical, piezoelectric, and thermal, in relation to the transducing elements (Shruthi, Amitha, & Blessy Baby, 2014) (Table 5.3).

The biosensor system (Figs. 5.2 and 5.3) is made from a bioreceptor, transducer, and related hardware components such as processor, screen, and signal amplifier. The biosensor can help in recognizing certain present elements that other type of sensors may not be able to detect. One of the recent popular biosensor is single nanoparticle-based surface plasmon resonance (SPR).

## 5.1.3 Bioreceptors

A bioreceptor is a biological molecular species, such as protein, enzyme, and antibodies, which binds or interact with certain compound or specific analyte (for the purpose of recognition) as a part of a biosensor that can generate an effect detected by the transducer (Figs. 5.3 and 5.4).

**Table 5.3** A brief history of bioelectronic noses.

| Year | Object of the invention |
|------|-------------------------|
| 1998 | Concept of bioelectronic nose |
| 1999 | A piezoelectric electrode used in the immobilization of a crude bullfrog cilia as a signal transducer |
| 2006 | SPR system to characterize molecular interaction between olfactory receptor and its cognate odor molecule |
| 2005 | Whole cell-based quartz crystal microbalance sensor system for selective recognition of odorant molecules |
| 2006 | A crude membrane expressing an olfactory protein was used for measuring odorants using a quartz crystal microbalance |
| 2011 | Biomimetic chemical sensors using nanoelectronic read out of olfactory receptor proteins |
| 2012 | Ultrasensitive flexible graphene based field-effect transistor (FET)-type bioelectronic nose |
| 2012 | Nanovesicle-based bioelectronic nose platform mimicking human olfactory signal transduction |
| 2013 | Peptide receptor-based bioelectronic nose for the real-time measurement |
| 2014 | Odorant detection using liposome containing olfactory receptor in the SPR system |
| 2014 | Olfactory biosensor using odorant-binding proteins from honeybee |
| 2015 | A surface acoustic wave bioelectronic nose for detection of volatile odorant molecules |

*Source*: Adapted and edited from Dung, T.T., Oh, Y., Choi, S.J., Kim, I.D., Oh, M.K., & Kim, M. (2018). Applications and advances in bioelectronic noses for odor sensing. *Sensors (Basel), 18*(1), 103. DOI: 10.3390/s18010103. PMID: 29301263; PMCID: PMC5795383. https://pubmed.ncbi.nlm.nih.gov/29301263/ (Dung et al., 2018).

**Figure 5.2**
Schematic
diagram of
biosensor
components.
*Redrawn and edited
from Kawamura, A.,
& Miyata, T. (2016).
Biosensors—
Section 4.2,
biomaterials
nanoarchitectonics.
Elsevier Inc.
(Kawamura and
Miyata, 2016).*

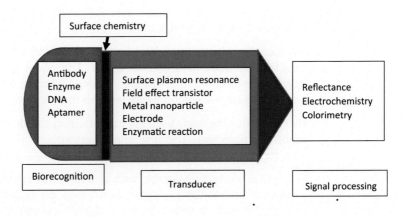

**Figure 5.3**
Different forms of
bioactive sensing
elements in a
biosensor. *Redrawn
and adapted from
Pethig, R., & Smith, S.
(2013). Introductory
bioelectronics—For
engineers and
physical scientists.
Wiley.*

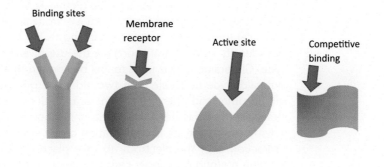

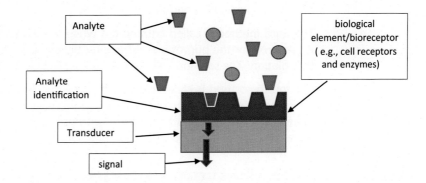

**Figure 5.4**
Illustration of
biosensor function
and components.

## 5.1.4 Biosensors construction

Many of the biosensors are made using screen printing. The substrate which
can be from plastic materials is coated with conducting polymer. There will be

an enzyme attached, for example, which is commonly used in constructing biosensors with two electrodes.

The most important part in the process of making a biosensor is the biological detection agent needed for the sensor. This agent can be part of a tissue or individual molecules, for example, enzymes (glucose oxidase). First the enzyme from biological materials is extracted and then purified.

There are four methods used for immobilizing the detection agent (biosensing agents/biomolecule). The first one is adsorption. This is one of the simplest ways for immobilization, and it is divided into physical and chemical methods.

1. Physical methods: Materials such as alumina or graphite powder for the purpose of immobilizing enzymes (anchoring the enzyme close to electrode) or materials such as carbon paste or alginate gel membrane for microorganisms are used. The next step is to use a cellophane dialysis membrane to confine the detection agent on the electrode (Pethig & Smith, 2013). The bonding in this method is weak, and therefore the electrode last for short period of time.
2. Chemical methods: Enzyme (crosslinking) is linked to a passive protein with high density (passive transport) via the use of a chemical called glutaraldehyde. Glutaraldehyde is used to link bovine serum albumin to enzymes. The glutaraldehyde molecule with two aldehyde groups will form a gel (which will become solid after the solvent evaporation) and is made up of protein molecules that has been linked via imine bonds. By using electrochemical polymerization for the immobilization of the enzyme, it will allow electronic interaction with the electrode to take place (Pethig & Smith, 2013). The bonding method is stronger compared with the physical method.

The second method is called covalent bonding which require low temperature and low ionic strength for the reaction to take place. However, this bonding method is between a functional group such as $-SH$, $NH_2$, $-COOH$, $-OH$, and $-C_6H_4OH$.

The third method is called entrapment where a solution of polymeric materials is prepared for the purpose of entrapping biological materials contained within the solution. The solution will be used to coat the electrode.

The fourth method is called crosslinking. This method involves the bonding of molecules (two or more) via covalent bonds. Using crosslinking agent the biomaterials will be bonded to it for the purpose of enforcing the attachment as glutaraldehyde bind these biological materials together.

## 5.2 System design

As mentioned previously, combining two different types of sensors, a biosensor and a technical sensor, can provide a wider range of detection with more focused outputs. However, this can be only done when a computing system software (including the hardware design) compiled for these two types of different sensors have been specifically coded for managing the outputs in conjunction with the two sensors, that is, there are three major parts of the device, such as a sample delivery system, a detection system, and a computing system, and each is considered individually and subdivided into a number of fully tested components.

## 5.2.1 Features and requirements

Regarding the number of sensors for the new MOD the decision was made to have two active sensors plus two reference active sensors—with the backup of additional sensors array. This kind of approach was made because each sensor differs in their functionality and characteristics, as well as there is no single sensor that can detect all kinds of odors/VOCs. The decision therefore is to employ several sensors, rather than just one sensor. In addition, as mentioned earlier, the device provided a backup of additional 10 sensors for the purpose of reducing errors, for the use on a variety of applications, and to provide additional information that the user may need.

The new portable MOD mobile unit system will be designed for validation within the laboratory environment and outside it, as well as for commercial and noncommercial applications, as previously discussed. Therefore the first sensor, for example, MMOS sensor, which can be used is p-type, and the second sensor is acoustic wave biosensors or single nanoparticle-based surface plasmon resonance (SPR).

Issues connected to the sensors that can be affected by humidity, temperature, sample and sample flow rate should be examined and tested before the actual design completed. This means all the devices within the system should be working efficiently to get the correct reading generated by the sensors. A humidity sensor and a temperature sensor are provided within the system (within each chamber). Moreover, the sensor chamber must be fully sealed (Box 5.1 and Box 5.2).

---

### Box 5.1 The sensor chambers

The sensor chambers were designed with one inlet and one outlet. The first chamber is the basic square chamber with two extensions for inlet and outlet of the samples. This shape is chosen in concordance with the form factor of the sensor array and the electronic nose device. The main aim was to minimize the volume of the chamber so as to increase the density of the sample within the chamber in minimum possible time. The second design is a modification over the first design where the sample gas follows a guided path over the sensor elements. A comb-like arrangement is made between the sensor columns to make a zigzag path for the sample gas. The third modification is inspired from biological olfaction and is targeted to create vortex around the sensor elements. Curved projections are made on the comb teeth to provide objection to smooth flow over the sensor elements and create a desired turbulent motion.

(Chowdhury, Kumar, Kumar, Karar, & Bhondekar, 2016).

---

## Box 5.2 eNose

The sample chamber serves as a place to put samples, and this chamber is made to isolate volatile organic compounds that are generated by samples from the influence of the surrounding environment. The sensor array is a core component of the eNose device, and this component serves to absorb and convert volatile compounds into electrical signals. Gas sensor output data in the form of an electric voltage ($V$) or electrical resistance ($R$) are then processed by the signal conditioning system to produce only information that is really useful for the pattern recognition system. Moreover, the pattern recognition system serves to predict, identify, and classify odor by comparing new odor data with previously recognized odor data. The eNose device does not work by searching for a particular gas molecule or compound but rather by looking for a unique pattern such as "fingerprint" from the analyzed air. This unique pattern is obtained using a chemical sensor array. The sensor array consists of several types of chemical sensors so that, when exposed to the aroma, each sensor in the array has a special response. For example, odorant A may produce a high response in one sensor and lower responses in others, while odorant B might produce high readings for sensors other than the one that "took" to odorant A. These different responses are stored in the database, which is then used to train pattern recognition algorithms in odor identification and classification.

(Subandri & Sarno, 2019).

The MOD hardware structure location identifier sensor, for example, GPS can be added as well to the system. This is already available in the market, such as GPS-622F board by RF Solutions Ltd., as it will provide further important information to the system (Gongora, Monroy, & Gonzalez-Jimenez, 2017). The four chambers contain four sensors, and the four sensors can be activated to function at the same time, whenever the need arise for additional analysis as well as additional 10 sensors can be activated, if needed, to make the overall total of 17 sensors. In addition to the abovementioned flexibility, it is possible to change a sensor (or sensors) with a different type of sensor if needs arise. However, when trying to analyze a sample for a specific element only, then it is enough to use the main two active sensors to obtain the required result. The main hardware parts of the system such as sample delivery system, detection system, and the computing system, each part has been considered in detail throughout the design of the new MOD system. The following diagrams illustrate the design of the system showing the basic principles of a device, such as the design provided in this chapter, which will function as a mobile MOD system (Figs. 5.5–5.14).

Concerning this design, the most important part in the device is the chamber. For this reason the chamber itself will be considered in more details.

## Introduction to Machine Olfaction Devices

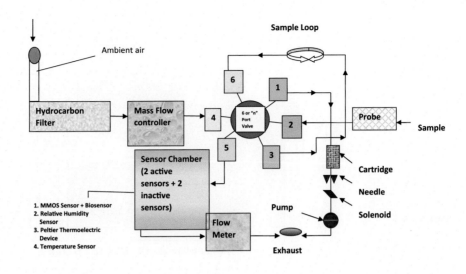

**Figure 5.5**
Schematic diagram of a system with two active sensors.

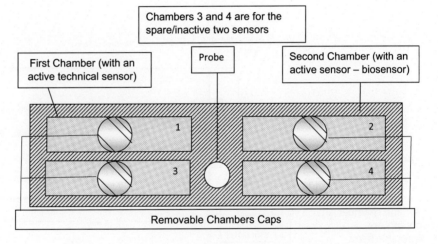

**Figure 5.6**
Top view of the proposed mobile MOD. *MOD*, Machine olfaction device.

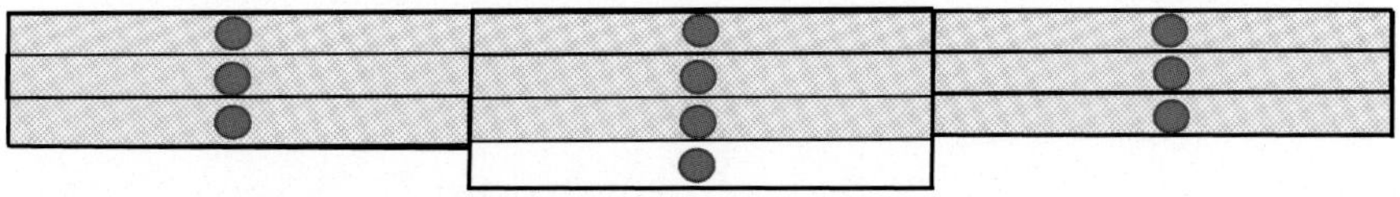

**Figure 5.7**
Four chambers with two active sensors and two reference active sensors.

The idea is that the content of the chamber will be a disposable part in the device. When the life cycle of the sensor is over, then the content of the chamber will be replaced with a new one.

The chamber is a permanently sealed component, which contains the sensor (MMOS or a biosensor), temperature sensor, relative humidity sensor, and the gas (sample) input and output paths.

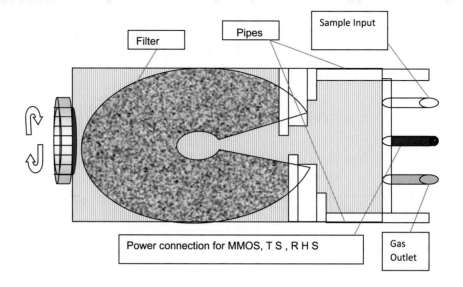

**Figure 5.8**
Additional 10 chambers (with sensors) which can be activated.

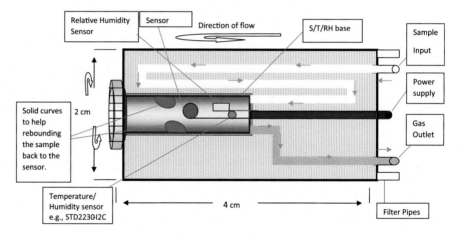

**Figure 5.9**
Filter attached at the back of the chamber—both filter and chamber contents are disposable items.

**Figure 5.10**
Cross-section of the chamber.

Power supply is made-up from:

1. Battery charger, 3.7 VDC (e.g.,Li-ion battery)
2. Converter (e.g.,3.3 V dc-dc converter)
3. Low drop out regulator (e.g.,1.8 V)

There is an additional space for another three chambers which can be used for at least two purposes—the first purpose is to work as a spare chamber which one of them can take over if the first chamber malfunctions, and the second purpose is for that the MOD which can be programmed to work with two sensors at the same time.

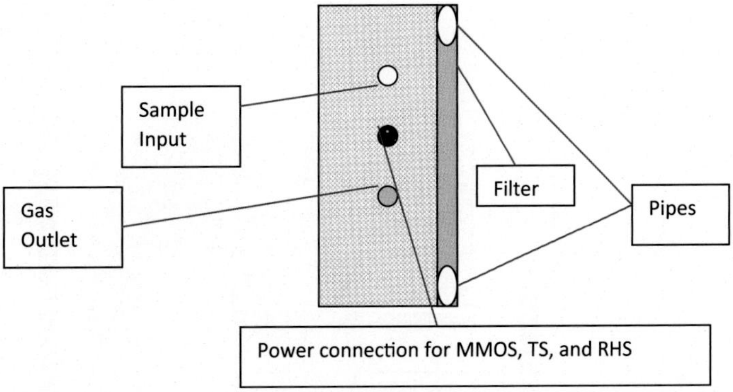

**Figure 5.11**
The outer bottom end of the chamber showing power connection, sample input, and outlet.

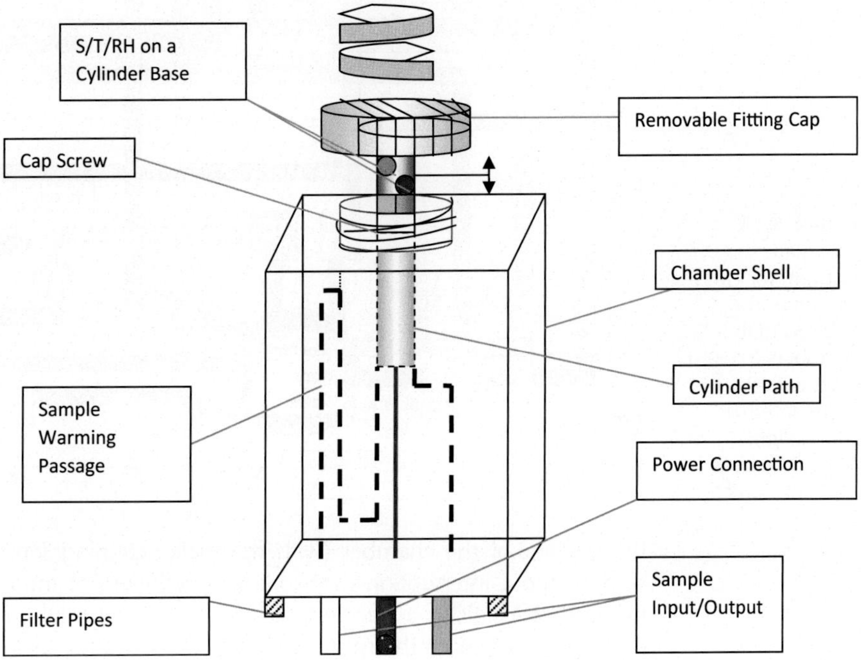

Sample Input

Gas Outlet

Filter

Pipes

Power connection for MMOS, TS, and RHS

**Figure 5.12**
The outer shell of the chamber showing the cylinder base (attached to it "from inside" the sensor/RH and temperature controller) being inserted inside the shell.

S/T/RH on a Cylinder Base

Cap Screw

Removable Fitting Cap

Chamber Shell

Cylinder Path

Sample Warming Passage

Power Connection

Filter Pipes

Sample Input/Output

For the purpose of our application the MOD need to work within an environment where a continuous monitoring is required, that is, 24 hours a day, over a period of many months. For this reason, to overcome the three-month life cycle of the sensor, a second chamber is provided for this purpose for the same type of sensor. That means, few days before the expiry date of the first sensor, the software program of the device will trigger the power supply for the second sensor. In this way the time required to stabilize the second sensor would have been achieved before the expiry date of the first sensor. which means the device function will not be interrupted as it usually happens if we have a single chamber. In the meantime

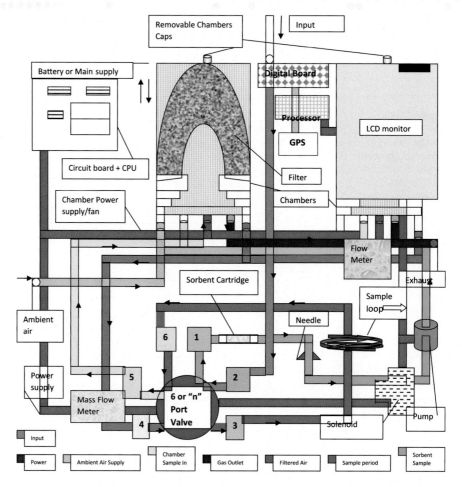

**Figure 5.13**
Schematic diagram for cross-section of the MOD with four chambers (plus) -i fixed with the rest of the internal connections. *MOD*, Machine olfaction device.

the content of the chamber (with the malfunctioning sensor) can be taken out safely without interrupting the work or functioning of the device, that is, for the purpose of installing a new sensor. The traditional method used is to open the chamber and replace the malfunctioning sensor with a new one and then wait for a period of time, which can be, in certain situations, up to 3 days to get the sensor "stabilization" back to normal. When functionality required and time is important, then the abovementioned method is no longer practical here. Wasting time and not being able to use the device are simply not acceptable in today's demanding market. Therefore the principles used in the new design can easily overcome such problems and difficulties. As we are aware, for any device to be successful on the market, the "market conditions" for this kind of device should be fulfilled. These conditions will be in the form of ease of use, continuous functionality (over a reasonable period of time) low cost, and, finally (most important point), the device is able to do the job intended for accurately.

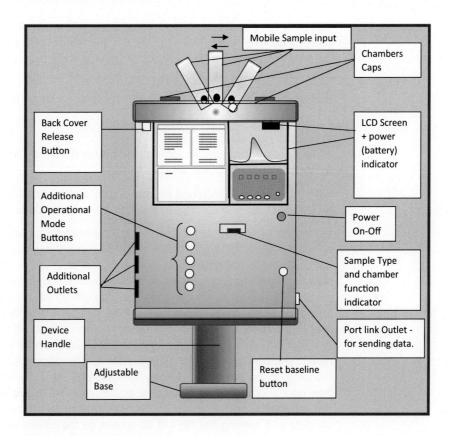

Mobile Sample input

Chambers
Caps

Back Cover
Release
Button

LCD Screen
+ power
(battery)
indicator

Additional
Operational
Mode
Buttons

Power
On-Off

Additional
Outlets

Sample Type
and chamber
function
indicator

Device
Handle

Port link Outlet -
for sending data.

Adjustable
Base

Reset baseline
button

**Figure 5.14**
A view of the new
MOD. *MOD,*
Machine olfaction
device.

## 5.2.2 MOD functions

The sample or the headspace is passed (or injected) into the sample port and via the n-port valve (or 6-port valve, depending on the design) and over the two active sensors. The two sensors responses outcome are passed via digital interface for control and data sampling of the sensor to the device processor for the analysis to be conducted.

The basic MOD operates in two different modes, the standby mode and sampling mode. The standby mode is for the purpose of stabilizing baseline of the sensor by allowing the sample flow (air or gas) to pass over the sensor as the device in this mode, that is, power switched on. The sampling mode, as the device is designed to function on a 24-hour basis, 7 days a week, then the device is ready for sampling. However, if the device is switched off, then around 30—60 seconds required for the purpose of the device to adjust and sensors to stabilize to the new flow rate. Few seconds are needed if there is a need to inject a sample. After obtaining the result from the injected sample, system clean-up will take over and may last 30 seconds (minimum) to 2 minutes (maximum), depending on the sample type and concentration. The n-port valve function enables the device, via the programmable switching, for the sample flow to occur between the standby mode and the sampling mode. The n-port valve can be assigned to different positions, according to the MOD design type. The option of standby mode or sampling mode can be done manually or automatically.

Real-time data acquisition will provide information about the following:

1. the flow rate in and out of the system;
2. relative humidity;
3. sensors' information, including real-time and average sensor resistance; and
4. temperature inside the sensor chambers.

When the device is switched on, specifically for the first time, the sensors must be left to stabilize, according to the instructions provided.

On the front panel of the MOD (main control program), certain buttons should be pressed to begin the device function and to check the device progress, such as for the display of sensor chamber temperature, real-time values, and resistance values, standby input flow, and standby flow rate. Certain button or buttons when pressed will display another window showing certain values and charts for selected functions. Some of the charts will display the data by showing past operations over a period, including sensors' vital information, which will help the user for further monitoring of the system, if needed. Flow rate can be verified against a flow meter. The configured sensor temperature can be done automatically (default) or manually set to the required level, for example, 30°C. By allocating the desired flow rates and time and pressing the button to start the sequence, the device will request confirmation to proceed. After confirmation, allocation for saving the sample data and naming the file will be shown on the screen. As soon as the above-mentioned confirmation provided, preparation function begins with data logging at various levels of the sample stages, showing the time for each step. After the completion of each stage, values are provided on the panel with the option to print a report or send to another location. Request for another sample to be tested will show on the screen, after clean-up stage has been completed. At the same time the device will return to standby mode for the flow rate and valve, that is, to the original condition prior to the sample test.

## References

Aliofkhazraei, M. (2014). *13.10. Recent developments in miniaturization of sensor technologies and their applications, Comprehensive Materials Processing* (13, pp. 245−306). https://www.sciencedirect.com/science/article/pii/B9780080965321013091.

Bhalla, N., Jolly, P., Formisano, N., & Estrela, P. (2016). Introduction to biosensors. *Essays in Biochemistry, 60*(2016), 1−8. Available from https://doi.org/10.1042/EBC20150001. <https://www.ncbi.nlm.nih.gov/pmc/articles/PMC4986445/pdf/bse0600001.pdf>.

Chowdhury, S., Kumar, D., Kumar, R., Karar, V., & Bhondekar, A. (2016) Simple sensor chamber design for hand-held electronic nose. <https://www.researchgate.net/publication/301204644_Simple_Sensor_Chamber_Design_For_Hand-Held_Electronic_Nose>.

Dung, T. T., Oh, Y., Choi, S. J., Kim, I. D., Oh, M. K., & Kim, M. (2018). Applications and advances in bioelectronic noses for odour sensing. *Sensors (Basel)*, *18*(1), 103. Available from https://doi.org/10.3390/s18010103. PMID: 29301263; PMCID: PMC5795383 <https://pubmed.ncbi.nlm.nih.gov/29301263/>.

Electronics Hub. (2019). *What are biosensors? principle, working, types and applications.* Retrieved from https://www.electronicshub.org/types-of-biosensors/. (Accessed 27 October 2020).

Gongora, A., Monroy, J., & Gonzalez-Jimenez, J. (2017). *An electronic architecture for multipurpose artificial noses.* Retrieved from https://www.hindawi.com/journals/js/2018/5427693/. (Accessed 31 October 2020).

Kawamura, A., & Miyata, T. (2016). *Biosensors—Section 4.2, biomaterials nanoarchitectonics.* Elsevier Inc.

Malhotra, B., & Pandey, C. (2017). *Biosensors: Fundamentals and applications.* Smithers Rapra Technology Ltd.

Nakamoto, T. (2016). *Essential of machine olfaction and test.* Wiley, Chapter Two. ISBN: 9781118768488.

Pethig, R., & Smith, S. (2013). *Introductory bioelectronics—For engineers and physical scientists.* Wiley.

Shruthi, G. S., Amitha, C. V., & Blessy Baby, M. (2014). Biosensors: A modern day achievement. *Journal of Instrumentation Technology*, *2*(1), 26–39.

Subandri, M., & Sarno, R. (2019). E-nose sensor array optimization based on volatile compound concentration data. *Journal of Physics: Conference Series.* <https://iopscience.iop.org/article/10.1088/1742-6596/1201/1/012003/pdf>.

# Microsystem

For more than 20 years, the designing and development of a micro- or portable electronic nose system were and still an essential requirement "because of its advantages such as small size, low cost, and easy manipulation in comparison with the ordinary electronic nose system" (Hong et al., 2000).

For the above reason, the idea behind the concept of "micro total-analysis system (μTAS)" introduced by Mantz (and coworkers) in (Manz, Graber, Widmer & 1990) can be incorporated into a mobile miniature machine olfaction devices (MOD).

The μTAS created huge interests among scientists and manufacturers because the conventional approach for chemical analysis is no longer sufficient in meeting the requirements of various applications. "μTAS offers a significant decrease in costs by dramatically reducing the volume of samples and reagents that are needed to perform a chemical analysis" (Geschke, Klank, & Tellemann, 2004).

Therefore when designing a MOD system, one of the most important question arising is "What is needed for the sensors to function accurately and repeatedly over longer periods of time?" This may also require the development of a smaller mobile unit, that is, a microlevel unit, which relies upon an acceptable stabilization time of the sensors.

The microlevel scale is the specification needed for the internal design of the new MOD, that is, the requirement and main objective is the internal design on a specified external miniature scale size for the device. Obviously, for any mobile device, and for devices with a selected MMOS and biosensor working at the same time, the need for low power consumption is essential.

## 6.1 System requirements

The following factors should be remembered as the minimum requirements needed for an MOD basic design. These important factors if not regulated or designed specifically for the purpose needed, it can be sources of errors impacting on the overall results:

1. operating temperature;
2. variations in humidity;

Introduction to Machine Olfaction Devices.
DOI: https://doi.org/10.1016/B978-0-12-822420-5.00002-7

3. the reference gas;
4. the flow rate of reference gas;
5. ambient air (e.g., using filter without the need for a pressure differential to help in passing the sample through the system);
6. chamber temperature control, using a thermoelectric controller;
7. flow control;
8. communication and control for each part of the device;
9. power (e.g., battery/light cells) and microelectric power for the sensors; and
10. software.

Parts of this system are required to be designed and manufactured to micro-scale level, which means that the content of the sensor chamber will be fabricated on a substrate (e.g., alumina), possibly using a screen printing technique.

## 6.2 Screen printing techniques (e.g., MMOS sensor/acoustic wave biosensors)

The following are the examples of techniques of the sensor-processing steps:

1. preparing/cleaning substrate;
2. electrode and heater formation;
3. sintering/heating to below the melting point of the substance powder [it depends on the type of materials chosen for the purpose of making a paste, which can be silicone elastomer polydimethylsiloxane (PDMS) or poly(methyl methacrylate) (PMMA), i.e., powder mixing—then hydrates, carbonates, or other compounds are decomposed, and the volatile material is expelled];
4. paste formation;
5. sensing layer printing; and
6. point wire bonding

The present design of the MMOS sensor is that when the power is switched on, the sensor requires time (from few hours up to 24 hours) to enable it to reach a stable baseline value; this is referred to by the sensor manufacturer as the "burn in time." The solution usually adopted to overcome this problem is to keep the power switched on all the time. This may not be a beneficial way in operating a mobile miniature MOD, as power needs to be conserved and used only during an operational time. Therefore a new way is needed, by switching off the power—apart, of course, either from the circuit connected to the sensor—or from the chamber containing the sensor. Otherwise, a non-metal-oxide semiconductor sensor (e.g., polymer) should be used instead.

To enhance system accuracy, four sensors will be active at the same time, that is, two MMOS sensors and two biosensors (e.g., two acoustic wave biosensors), while two additional sensors similar, to the four sensors, called "the final" will be activated if there is no match among the active four sensors regarding result. This takes place only if the difference between the results is noticeable. On the other hand, if the difference is very small, then the outcome will be produced by taking the average result obtained from the four sensors and, consequently, there will be no need to consult the final sensor. To explain the abovementioned procedure in more details, it is useful to explain each step as follows:

A MMOS sensor and a biosensor will perform the sample analysis first, and the result will be crosschecked by comparing with the result obtained from the second MMOS and second biosensor already active within the same system. If it matched or is very close in matching, then the result will be passed or the average is taken and passed as accurate.

If it did not match, then the same first result (i.e., from the first MMOS and the first biosensor) will be checked in comparison with the third MMOS sensor, by then they would have been activated, as well as with the third biosensor, that is, the final. If it matched or is very close in matching, then it will be passed or the average taken and will be passed as accurate.

On the other hand, if there is no match among one or more of the four sensors, then either the software indicates which sensor is generating a different result than the rest of the sensors and how it may provide possible solution, including repeating the test on the same sensor, or, if the error persists, the device indicates the need to replace the nonmatching sensor. If the result is still in disagreement after replacing a possibly faulty sensor or when the error originates from more than one sensor, then the overall result is failed and the device will direct you to look for other factors which may be causing this failure, including the requirement to restart the system as a first simple solution.

The abovementioned method will remove the need for using established methods for comparison (e.g., HSGC-MS analysis) as means of comparison.

## 6.3 Micro total-analysis system

Within the past 20 years or so, advancement in device miniaturization has led to the development of integrated microfluidic devices, μTAS; other researchers/manufacturers call it "lab-on-chip."

Having said that, there are a number of steps which need to be followed carefully *before* and *during* the process of designing such a system. As mentioned earlier, the first step will be the specifications of the device. Obviously, these specifications are related to the type of chemical analysis which the system will work on. To do that, a number of points should be considered (Geschke et al., 2004) such as the following:

a. reagents used;
b. reaction kinetics;

  c. reaction temperature;
  d. detection method;
  e. range of detection; and
  f. detection limit.

These, which can be applied to various MOD systems or electronic noses, are the basic similarities between an eNose and μTAS (i.e., nose on chip).

Finally, the materials which will be used to make the device are determined by the type of chemistry that is used or takes place within the system. Miniaturizing internal various different parts of the MOD system to microscale or nanoscale level, challenges will surface, such as friction begin to matter, as well as size effects (Box 6.1).

---

## Box 6.1 The size effects

The contributors to size effects can be put into three groups: density, shape, and structural effects. Density effects relate to the inhomogeneity of materials at a small scale. If the average grain size of a material is considered to be of the order of 40 mm, a macrosize component would contain millions of grains, meaning that the material could be modeled as homogeneous. Going to the microdomain, a 500-mm feature would only contain about 12 grains over the cross-section, resulting in the properties of the individual grains influencing the forming process. Shape effects are closely related to the surface-to-volume ratio of the component. Consider a dice with a scalable side length of $a$. The volume then scales with the third power $a^3$, whereas the surface area scales to the second power $a^2$, meaning that a small component will have a higher surface-to-volume ratio, by $a$, when compared with its larger equivalent. The surface-to-volume ratio is an important parameter when considering an ejection or handling situation, where the friction is a factor of the component surface area and the component strength is related to the material volume. Designing an ejector system to remove a microcomponent from the forming die is challenging due to the risk of collapse or reverse forming during ejection. Furthermore, small components are prone to the sticking effect, where adhesive forces between the gripper and the component outweigh the gravitational forces. These adhesive forces primarily consist of surface tensions, van der Waals forces, and electrostatic forces and can be the limiting factor of a handling system for microcomponents. The third effect contributing to the size effect is the group of microstructural effects. This group of effects is made of physical elements which either experience a physical length limitation, where it is not practical to scale the microgeometry, or where secondary scaling effects come into play.

Arentoft et al., 2010

# 6.4 Materials

There are a number of possible materials which can be used in the fabrication process to produce a μTAS, mostly polymers. Two of them are explained further.

1. *Silicone elastomer PDMS:* PDMS is a good example of excellent materials used in microfabrications; for example, it was used in constructing microchannel systems to be used with biological samples in aqueous solutions (Zhou & Xu, 2001). It can give excellent results in replication (molding), it is a transparent material (down to 280 nm, nontoxic, and flexible) and can be used within a wide temperature range with a low chemical reactivity (Löttersy, Olthuis, Veltink, & Bergveld, 1997).

2. *Poly(methyl methacrylate):* PMMA, a synthetic resin belonging to the family of polymeric organic compounds, is manufactured by bulk, solution, suspension, or emulsion polymerization of MMA monomer, usually in aqueous suspension. The resin has excellent transparency, high surface hardness, and coloring property. It is used as a replacement for glass. Its application includes the semifinished products of automotive components, electronic parts, illuminated display sheet for advertising, building materials and liquid crystal display plates.

# 6.5 Microfabrications

The process of miniaturization and microfabrication are an important part of science and engineering, as there are many advantages to be gained by converging macrodevices into miniformat or microformat. As we know, miniaturization is usually described as the scaling down of macrodevices. On the other hand, microfabrication involves the application of an entirely new set of concepts to achieve the desired goals (Xie, Ramanathan, & Danielsson, 2000).

The microfabrication technologies make devices ranging from few millimeters to few microns, producing devices (or part of devices) such as motors, pivots, linkages, and other similar mechanical and nonmechanical devices.

Technologies such as silicon surface micromachining which is used for the purpose of making microsensors, involve the removing of silicon substance by implementing a method called "wet chemical" or "dry plasma" for the purpose of constructing a three-dimensional microstructure which can be either from silicon or from other type of materials, produces structures inside a substrate, whereas silicon bulk micromachining simply means the production of the structure inside the substrate.

At present, complementary metal oxide semiconductor technology is a good example of the efficiency of microfabrication technologies. Processing steps taken from semiconductor technology are used "in combination with dedicated micromachining steps" to fabricate three-dimensional parts for mechanical structures, as it eventually will form the basis for the chemical and biosensors (Hierlemann, Brand, Hagleitner, & Baltes, 2003).

According to Hierlemann et al. (2003), sensor characteristics can be improved and/or it is possible to develop devices with new functionality by simply using microfabrication techniques in different ways. Obviously, this cannot be done using conventional fabrication technology.

## 6.5.1 Microfabrication steps

Basic microfabrication techniques are deposition, patterning, doping, and etching. A brief description for each step of the microfabrication techniques has been described further.

- *Deposition:* Usually a thin layer, for example, insulating silicon dioxide film, can be used as the "deposit" on a substrate.
- *Pattering:* A sensitive photoresist layer can be deposited on top of the film and then patterned via photolithography.
- *Doping:* The purpose is to modify the electrical conductivity of semiconducting materials, for example, silicon. It is the main step in the fabricating process for semiconductor. Dopant atoms are passed by ion implantation or sometimes from gaseous, liquid, or solid diffusion.
- *Etching:* Two different types of etching can be used, wet etching and dry etching, depending on the exact requirements in the manufacturing process.

More details about these techniques will be discussed later on. But before going deeper into engineering aspects of this technology (Fig. 6.1), there are a few things that need to be looked at first, as follows(Geschke et al., 2004):

1. manufacturing environment (clean rooms);
2. theoretical aspects of microfluidics;
3. components of microfluidics; and
4. microsystems and eNose.

## 6.5.2 Packaging of microsystems

### 6.5.2.1 Manufacturing environment (clean rooms)

In μTAS, the dimensions of working components are down to micrometers, and therefore small particles within the surrounding environment can destroy the function of μTAS. Obviously, the purpose for using "clean rooms" is to prevent small particles and/or contamination taking place. The clean rooms therefore are an essential part of the high-tech manufacturing (and assembly) of sensitive material/devices.

These rooms can be small, very small chambers (microenvironments), or large-scale spacious rooms (ballrooms). The technology used in this kind of rooms are

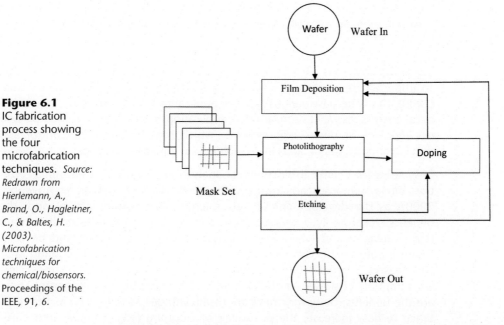

**Figure 6.1**
IC fabrication process showing the four microfabrication techniques. *Source: Redrawn from Hierlemann, A., Brand, O., Hagleitner, C., & Baltes, H. (2003). Microfabrication techniques for chemical/biosensors.* Proceedings of the IEEE, 91, 6.

variety types of industries such as semiconductor assembly, biotechnology, pharmaceutical, aerospace, medical devices, and particularly (and in general), the microsystem engineering, which is the main topic being discussed in this chapter.

A number of things will be taken into consideration, such as "the cleanliness class," "fabrication type," "ESD control," passages, and a gowning/changing area. The standard for each class is determined through the use of government specification (in the United States known as Federal Standard 209E) to provide a qualified and standardized method for measuring how clean the air is in a clean room. There are six classes compiled to indicate the cleanliness of the clean room.

Particles bigger than one-half of a micron in one cubic foot of clean room air determine the class number. For example, 100 particles (the size of each particle bigger than half microns) in one cubic foot of air are classified as Class 100 (Geschke et al., 2004).

According to the US standard classification mentioned earlier, the six classes are named as follows:

1. Class 1 (ISO 3);
2. Class 10 (ISO 4);
3. Class 100 (ISO 5);
4. Class 1,000 (ISO 6);
5. Class 10,000 (ISO 7); and
6. Class 100,000 (ISO 8).

There are five main fabrication styles for clean rooms: conventional, modular hardwall, modular softwall, minienvironment, and microenvironment.

## 6.5.2.2 Theoretical aspects of microfluidics

Currently, at various parts of the world, the microfluidic systems have become an important research topic. According to Nguyen, Meng, Black, and White (2000), this is because various types of applications in microbiology, microchemical analysis, as well as similar other applications encourage more designs and development of microfluidic systems, which has the ability to control and deliver various types of fluids in "nanoliter" to "microliter" ranges.

In systems for chemical/biochemical analysis and synthesis, the miniaturization offers numerous advantages (Rife, Bell, Horwitz, Kabler, & Auyeung, 2000): as the development of new microfluidics instruments expand, the benefits, accordingly, increase. These benefits can be manifested in a number of areas as and when the scale gets closer to microlevel and beyond. Examples of benefits can be seen in the increased reaction and cooling rates (the reason is larger relative surface areas and smaller diffusion times), less consumption of power, mobility, and smaller quantities of reagents. These are the benefits; however, there are disadvantages as well, such as the resistance to flow increase, the possibility of clogging (e.g., bubbles and particles), and difficulties in the laminar flow regime when there is a need for mixing (Rife et al., 2000).

There are major points in microfluidics, which should be looked at carefully when designing a device (Geschke et al., 2004) such as the following:

1. transport process;
2. system design; and
3. application
1. *Transport process:* To look at the transport process, we need to look at the type of transport being used. In the microfluidics, there are two types of transport:
   a. directed transport and
   b. statistical transport.

Directed transport is done and controlled by exerting "work" on the fluid itself. The work is produced by a pump or induced electrically by a voltage, with each having its own name: the former is called "pressure-driven flow" and the latter "electroosmotic flow."

The other type, statistical transport (or diffusion), mainly happen between two liquids, one with high concentration of certain type of molecules and the other with zero or lower concentration. The transport (movement) continues until a balance has been established.

In many cases, there is a need for both methods to complete a successful method of transport. Therefore the electroosmotic flow method of transport (with little help from statistical transport) will be used in constructing a miniature device (e.g., MOD) in relation to this work.

2. *System design*

   a. *Transport:* When designing a microsystem, it will be helpful to know Péclet number to help us with the flow injection analysis. Péclet number relates to the effectiveness of mass transport by advection to transport by either dispersion or diffusion:

$P = v_x d/D_d,$

where $P$ = Péclet number; $v_x$ = advective velocity; $d$ = average grain diameter; and $D_d$ = coefficient of diffusion.

In most microsystems, the speed of the flow (velocity) is slow; therefore the most important factor is the channel length, which is the variable that will determine the Péclet number. Consequently, if the Péclet number is known (i.e., for a microchannel and its dimension) (Geschke et al., 2004), then it will be much easier in designing the system; this is because we will be able to know when diffusion dominates the microfluidic flow when we find out that the Péclet number is smaller than one. However, if this is not the case, that is, the Péclet number is higher than 1, then the flow will occur by an external applied force, and therefore the diffusion here will have a minimal part to play in relation to the flow.

Laminar flow and diffusion are the main transport methods within the microfluidic system. Therefore to help in the designing of a functional microsystem the need to understand laminar flow and diffusion is an important part in relation to the microfluidic device (Fig. 6.2).

For example, if we take a T junction where two liquids meet, they will flow side by side in streams, that is, laminar flow, and eventually, they mix slowly, that is, diffusion (Geschke et al., 2004). Therefore the mixing happens (assuming there is no external force) simply because of the effect of diffusion. The diffusion may help in the device of this project to transport the sample across the

**Figure 6.2**
Example of laminar flow with turbulent flow above it. *Source: Adapted and redrawn from Ducrée, J., & Zengerle, R. (n.d.). Microfluidics. Institute of Microsystem Technology, University Freiburg.*

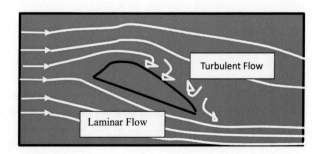

Turbulent Flow

Laminar Flow

tubes; however, this kind of transport is very slow, and therefore we will need to design a micropump for this purpose.

     **b.** *Micropump:* As the design of the MOD will be on a miniature/micro-scale level, a certain type of a "micropump" to help in passing the sample to the sensor(s) will be needed. For this reason a brief introduction of the micropump is given in this section; however, more details will be provided later on.

Various work and research have been done in different areas on flexural plate wave (FPW) devices for pumping fluids. But what is FPW? The names used for FPW are ultrasonic microfluidic FPW devices or magnetically excited FPW devices.

Acoustic FPWs are created from piezoelectric motion where a FPW oscillate at a single resonant frequency, and when liquid is loaded, its resonant frequency changes as a function of liquid density.

"Ultrasonically driven micromachined silicon needle. Originally designed for surgery, when its tip is immersed in a liquid and the needle is driven ultrasonically, the hollow shaft acts as a pump. It can also be used as an ultrasonic source for lysing cells." (BASC, undated) (Figs. 6.3 and 6.4).

There are a number of methods used in designing a micropump, with the most common one being the use of a membrane—the movement of the membrane itself will simulate the function of an ordinary piston pump. It usually consists of a pump chamber having a flexible diaphragm driven by an actuator (Cao, Mantell, & Polla, 2000). The fluid, or gas, will be drawn in when the membrane moves out, increasing the size of the chamber; therefore the pump is at a "suction or supply mode." When the reverse happen, that is, when the membrane moves in, the chamber size will be reduced and the gas or fluid will be pushed out, and the mode is therefore "pumping mode" (Cao et al., 2000) (Box 6.2).

In the Netherlands during 1988 a two passive-valve micropump was designed (Van Lintel, Van de Pol, & Bouwstra, 1988); however, there is another type of membrane micropump, the valveless diffuser/nozzle micropump (Cao et al., 2000) (Figs. 6.5, 6.6, and 6.7).

3. *Application:* There are many applications which can be applied using the microfluidic system. The biggest area where this system can be used is the health sector, as the microsystem can be used in areas such as diagnostic, monitoring and drug delivery, and, particularly, testing and handling blood (Geschke et al., 2004). In the MOD design, the use of microfluidic system, whether to deliver a liquid or a gas, will form an important part of the miniature MOD device. The device should be able to handle biological and non-biological samples at the same time.

**Figure 6.3**
Micromachining to fabricate an integrated CMOS cantilever with on-chip circuitry. (A) Chip after completion of the CMOS process and subsequent patterning of the silicon nitride etching mask. Micromachining steps: (B) anisotropic silicon etching and (C) frontside reactive ion etching. *Source: Adapted and redrawn from Hierlemann, A., Brand, O., Hagleitner, C., & Baltes, H. (2003). Microfabrication techniques for chemical/biosensors. Proceedings of the IEEE, 91, 6.*

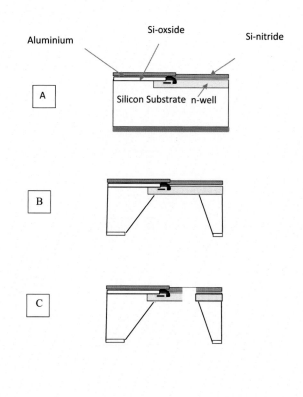

### 6.5.2.3 Components of microfluidics

Some of the components have been already mentioned briefly, such as micropumps and valves; these components are the most important elements in a microfluidic system. However, there are other parts which need to be looked at within the system, such as the following:

1. injecting;
2. measuring temperature within the system; and
3. electrochemical sensors.

1. *Injecting:* The ability to input a very small sample of liquid (in the MOD, a gas sample), repeatedly on a daily basis and with no difficulty, is vital for the success of any microsystem. For reasons related to various analytical techniques, it may be necessary to inject a defined/measured amount of sample. This "measured" sampling is usually referred to as "aliquoting" (Geschke et al., 2004).
2. *Measuring temperature within the system:* In many cases, it is important to measure the temperature of the sample within the microsystem. The reason for this is because the kind of system usually used for the production

**Figure 6.4**
Examples of basic operational modes of a micromechanical cantilever−based sensor. (A) In the thermogravimeter application the cantilever is excited to vibrate at its resonance frequency, which is determined by the mass of an attached specimen. (B) The calorimeter device can measure permanent and transient temperature variations via the bimetallic effect by detecting the bending of the cantilever beam. (C) Surface stress changes arising from specific molecular adsorption on one side cause a permanent bending of the cantilever beam (Berger et al., 1997). *Source: Adapted and redrawn from Berger, R., Lang, H. P., Ramseyer, J. P., Battiston, F., Fabian, J. H., Scandella, L., ... Gimzewskia, J. K. (1997). Transduction principles and integration of chemical sensors into a micromechanicl array device. Physical Sciences, 9, RZ 2986 (#93032).*

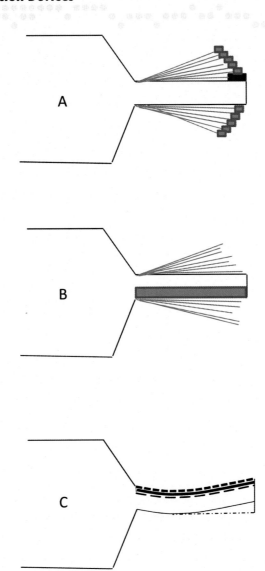

## Box 6.2 Active microvalves

An active microvalve consists of a device body that contains the fluid under pressure, a valve seat to modify the fluid flow, and an actuator to control the position of the valve seat. The first-reported microvalve was designed as an injection valve for use in integrated gas chromatography. It had a silicon valve seat and a nickel diaphragm actuated by an external solenoid. Following this first design a large number of microvalves have been designed and reported, and they can be classified on the basis of the actuation method employed. These methods include pneumatic, thermopneumatic, thermomechanic, piezoelectric, electrostatic, electromagnetic, electrochemical, and capillary force microvalves.

Pneumatic valves have a membrane structure as the valve seat. Although pneumatic actuation is based on a very simple principle, it requires an external pressure source which makes the pneumatic valves unsuitable for most compact applications. A low spring constant is also an important parameter, and to achieve it, thin membranes or corrugated membranes have to be designed. Soft elastic materials, such as silicon rubber or Parylene, can be used to realize the low spring constant, whereas hard materials such as silicon and glass are problematic. Thermopneumatic valves utilize a sealed pressure cavity filled with a liquid. Actuation is based on the change in the volume of the sealed liquid. The phase change from liquid to gas or from solid to liquid can also be used if a larger volume expansion is required. The disadvantage of these types of valves is the incompatibility of the technology because the liquid has to be primed, filled, and sealed individually. Thin films of solid paraffin material could be used as an alternative as they could be integrated in the batch fabrication. (Gardner & Cole, 2002).

(microreactor) or for analysis of chemical and biological reactions which make the scale of temperature and how to measure it forms another important part of the system (Geschke et al., 2004).

3. *Electrochemical sensors:* The principles of measurement in this sensors are "potentiometry" and "amperometry," an indication that the current flow between the two electrodes is constant (i.e., the potential difference is constant between the two electrodes). Electrochemical sensors in general can be easily fabricated as the need is for two electrodes made from metals such as gold, silver, platinum, or graphite and, for example, electrochemical sensors to be applied if the sample is turbid. There are a number of ways in miniaturizing sensors, and these will be discussed in details later on.

**Figure 6.5**
Electron micrograph of fabricated microvalve (Bien, Mitchell, & Gamble, 2003). *Source: Adapted from Bien D. C. S., Mitchell S. J. N., & Gamble, H. S. (2003). Fabrication and characterization of a micromachined passive valve.* Journal of Micromechanics and Microengineering, *13, 557–562. PII: S0960–1317(03) 55583–8.*

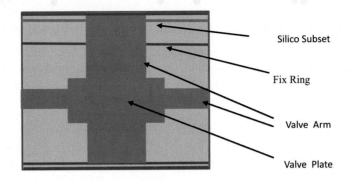

Silico Subset

Fix Ring

Valve Arm

Valve Plate

**Figure 6.6**
Example of microdisplacement pumps valve (cantilever-passive valve), showing two different positions, supply, and pump mode. *Source: Adapted from Ducrée, J., & Zengerle, R. (n.d.). Microfluidics.* Institute of Microsystem Technology, University Freiburg.

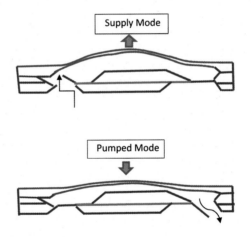

Supply Mode

Pumped Mode

### 6.5.2.4 Microsystems and machine olfaction devices

Many names are being used currently for the system described, for example, "biosystem-on-a-chip", "eNose on-a-chip," and, to detect dangerous or illicit substances, "dog-on-chip." The eNose designed using microsystem techniques is simply a computer chip with sensors that can detect various types of chemicals, regardless whether the human nose is capable of "smelling" them or not.

**Figure 6.7**
Schematic diagram (cross-section) of valveless micropump (Pan, Ng, Wu, and Lee, 2003). *Source: Adapted from Pan, L. S., Ng, T. Y., Wu, X. H., & Lee, H. P. (2003). Analysis of valveless micropumps with inertial effects.* Journal of Micromechanics and Microengineering. *13, 390–399. PII: S0960–1317(03) 53076–5.*

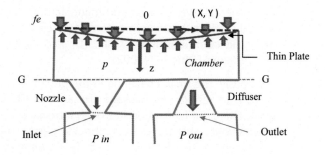

The main applications related to the "microfabricated" chemical systems, have been in the field of analytical chemistry, as this kind of chemistry is mainly concerned with the collection of "qualitative and quantitative" data for various types of samples (Geschke et al., 2004), so questions will be asked, such as:

1. For identification of elements/components, What does the sample contain?
2. For quantity, how much of the "element x" contained in the sample?
3. The form in which "element x" exists?
4. The method that "element x" entered the sample.

Using a microsystem analytical method, these questions should be answered in various steps, for example, sampling, with the various necessary steps associated with the preparation of the sample, detection, sensing, and evaluation.

Therefore the idea of designing μTAS will be the answer for many problems, not mentioning saving on cost, time, energy, and reducing waste.

### 6.5.2.5 Microsystems packaging

This is an important aspect for achieving a proper and efficient functioning of a microsystem device, as without the knowledge and understanding of the materials involved, system and subsystem behaviors, and the limitation of the present packaging technology, the final microsystem device may not work as intended in the first place (Geschke et al., 2004). Since microsystem packaging may not be an academic subject, the skills required in this area are generally lacking. "Today's microsystems packaging engineers also need deep, fundamental, system-level and manufacturing knowledge; an understanding of global markets, business, economics, foreign language and culture; and leadership and communications skills" (Dr Rao Tummala, President, CPMT Society, IEEE, Atlanta *EP&P*, 6/1/2002)

Obviously, the main challenge in the process of packaging within any microsystems is to reduce the package size and, in the process, to reduce the cost as well. Therefore fabrication has been introduced for this purpose: through-wafer interconnect structures. Interconnects permit device-scale packaging and are compatible with "flip chip" assembly. Bonding methods to join various components (and the driving circuitry) are usually achieved using a low temperature for this purpose.

According to Geschke et al. (2004), for a detailed investigation into this topic, the following need to be considered:

1. level of packaging;
2. design process in packaging;
3. influencing factors in packaging and design;
4. factors influencing package reliability;
5. interconnections; and
6. comparison of micromachining materials.

# References

Arentoft, M., Eriksen, R., & Hansen, H. (2010). Micro-bulk-forming. In Y. Qin (Ed.), *Micromanufacturing engineering and technology* (p. 115).

BASC (Berkeley Sensor & Actuator Center). (Undated). University of California Berkeley Sensor & Actuator Center. Focused Flow Micropump Using Ultrasonic Flexural Plate Waves. Accessed 10.10.16. <https://vcresearch.berkeley.edu/research-unit/berkeley-sensor-actuator-center> Or: <https://ntnlab.com/wp-content/uploads/2016/11/J12JBMFocusedFPW.pdf>

Berger, R., Lang, H. P., Ramseyer, J. P., Battiston, F., Fabian, J. H., Scandella, L., ... Gimzewskia, J. K. (1997). Transduction principles and integration of chemical sensors into a micromechanicl array device. *Physical Sciences, 9*, RZ 2986 (# 93032).

Bien, D. C. S., Mitchell, S. J. N., & Gamble, H. S. (2003). Fabrication and characterization of a micromachined passive valve. *Journal of Micromechanics and Microengineering, 13*, 557–562, PII: S0960–1317(03)55583–8.

Cao, L., Mantell, S., & Polla, D. (2000). Design and simulation of an implantable medical drug delivery system using microelectromechanical systems technology. *Sensors and Actuators A Physical, 94*(2001), 117–125.

Gardner, J., & Cole, M. (2002). Integrated electronic noses and microsystems for chemical analysis. In T. Pearce, S. Schiffman, & J. Gardner (Eds.), *Handbook of machine olfaction*. Wiley. Print ISBN:9783527303588; Online ISBN:9783527601592; Available from https://doi.org/10.1002/3527601597. <https://onlinelibrary.wiley.com/doi/book/10.1002/3527601597>.

Geschke, O., Klank, H., & Tellemann, P. (2004). *Microsystem engineering of lab-on-a-chip devices*. Weinheim: Wiley-VCH Verlag GmbH & Co. KGaA, ISBN 3-527-30733-8.

Hierlemann, A., Brand, O., Hagleitner, C., & Baltes, H. (2003). Microfabrication techniques for chemical/biosensors. *Proceedings of the IEEE, Vol. 91*, 6.

Hong, H.-K., Kwon, C. H., Kim, S.-R., Yun, D. H., Lee, K., & Sung, Y. K. (2000). Portable electronic nose system with gas sensor array and artificial neural network. *Sensors and Actuators B Chemical, 66*(1–3), 49–52.

Löttersy, J. C., Olthuis, W., Veltink, P. H., & Bergveld, P. (1997). The mechanical properties of the rubber elastic polymer polydimethylsiloxane for sensor applications. *Journal of Micromechanics and Microengineering, 7*, 145–147.

Manz A., Graber N., Widmer H. (1990) Miniaturized total chemical analysis systems: A novel concept for chemical sensing. *Sensors and Actuators B: Chemical. 1*(1–6) January 1990 244–248. < https://www.sciencedirect.com/science/article/abs/pii/092540059080209I > .

Nguyen, N.-T., Meng, A. H., Black, J., & White, R. ,M. (2000). *Integrated flow sensor for* in situ *measurement and control of acoustic streaming in flexural plate wave micropumps. Sensors and Actuators, 79*, 115–121.

Pan, L. S., Ng, T. Y., Wu, X. H., & Lee, H. P. (2003). Analysis of valveless micropumps with inertial effects. *Journal of Micromechanics and Microengineering., 13*, 390–399, PII: S0960–1317(03)53076–5.

Rife, J. C., Bell, M. I., Horwitz, J. S., Kabler, M. N., & Auyeung, R. Y. W. J. (2000). Miniature valveless ultrasonic pumps and mixers. *Sensors and Actuators A Physical, 86*(1–2), 135–140.

Van Lintel, H. T. G., Van de Pol, F. C. M., & Bouwstra, A. (1988). A piezoelectric micropump based on micromachining of silicon. *Sensors and Actuators, 15*(1998), 153–167.

Xie, B., Ramanathan, K., & Danielsson, B. (2000). Mini/micro thermal biosensors and other related devices for biochemical/clinical analysis and monitoring. *Trends in Analytical Chemistry, 19*(5), 340–349.

Zhou, Y., & Xu, S. (Eds.). (2001). *International conference on sensor technology (ISTC 2001)* (Vol. 4414, pp. 38–39), October 10–12, 2001, Wuhan, China. ISBN 0–8194–4119–8.

# Nanoelectronic systems

Nanoelectronics is a specialized part of nanotechnology. In this chapter, we are going beyond traditional electronic circuits and deeper into molecular-scale technology, such as the fabrication of atomic wires, single-electron tunneling (SET) devices and atto-Farad structures. In other words, we are researching into and looking at the properties of devices which have dimensions at the nanometer scale (Box 7.1).

The different types of properties that nanoelectronics exhibit in this field are occasionally referred to by the name "disruptive technology."

## 7.1 What is nanoelectronics?

For existing integrated circuits (ICs), which are also known as *microelectronics*, the term *micro* derives from microfabrication technology, which embraces all highly sophisticated techniques like optical- and electron-beam lithography, metallization, implantation, and etching, which between them allow the generation of structures on the scale of 1 μm. Nanoelectronics, however, is concerned with understanding and exploiting the properties of devices that have dimensions at the nanometer scale.

In general terms, nanoelectronics will probably have two main origins, that is, to some extent, existing microelectronics will gradually evolve into nanoelectronics. In fact, this has already happened because the smallest feature size of present ICs is well below 1 μm ($10^{-6}$). It is currently accepted that optical lithography can be used for design rules down to at least about 100 nm ($1 \text{ nm} = 10^{-9}$). However, this would incur an increasing process and mask complexity and, consequently, increasing cost. There may also be further changes, for example, based on replacing or complementing today's techniques by devices based on using inorganic, organic, or biological molecules as electronic devices

> ## Box 7.1 Nanotechnology historical dates
>
> 1959 Richard Feynman lecture, "There is Plenty of Room at the Bottom."
>
> 1960s "Bottom-up" nanotechnology approach used in molecular structures.
>
> *(Continued)*

**Introduction to Machine Olfaction Devices.**
DOI: https://doi.org/10.1016/B978-0-12-822420-5.00014-3

> **Box 7.1 (Continued)**
>
> 1970s "Top-down" nanotechnology approach used by electrical engineers.
>
> 1981 The scanning tunneling microscope invented.
>
> 1981 The article by Drexler, "Molecular Engineering: An Approach to the Development of General Capabilities for Molecular Manipulation," published.
>
> 1985 A new form of carbon molecule discovered by Richard Smalley.
>
> 1986 Drexler's "*Engines of Creation*" book published.
>
> 1989 First international nanotechnology conference hold by Drexler's Foresight Institute.
>
> 1990 Famous IBM miniaturized logo.
>
> 1991 In Japan, Sumio Iijima discovered carbon nanotubes.
>
> 1992 The book "*Nanosystems: Molecular Machinery Manufacturing and computing*" by Drexler published.
>
> 1997 First molecular nanotechnology company "Zyvex" established.
>
> 1999 Mark Reed and James M. Tour design computer switch on molecular-scale (a single molecule).
>
> 2001 IBM develop a method to growing nanotubes.
>
> 2001 Scientists in Hong Kong proved that nanotubes can "superconduct."
>
> 2001 IBM build first logic gate (NOT) on a nanotube (1.4 nm) surface.
>
> 2003 First nanochip (65 nm) by Intel.
>
> 2005 Erik Winfree and Paul Rothemund developed theories for DNA-based computation and algorithmic self-assembly.
>
> 2010 Nanoscale assembly devices via the creation of DNA-robotic by Nadrian Seeman.

or components of electronic devices. This field is now known as "molecular electronics." Molecular-scale electronics has been widely touted as the next step in electronic miniaturization, with theory and research suggesting that single molecules may have the capability to take the place of today's much larger electronic components.

A further area under active investigation is spintronics, which has been defined as "the utilization of electron spin for significantly enhanced or fundamentally new device functionality," and can be considered as an atomic- or molecular-scale version for the use of magnetic storage media.

In this chapter, the question of "Why nanoelectronics for MOD?" will be looked at with some examples, but to understand the basic industrial methods for "electronic circuits" preparation, lithography has been looked at and then other aspects of nanoelectronics dealt with, such as nanotubes. Therefore, the focus is the implementation of nanotechnology, that is, nanoelectronic, as well as bionanotechnology, for the purpose of further improvement to the internal miniaturization of MOD systems, mainly not only for higher performance and cost reduction but also for the impeding purpose of various devices.

## 7.2 Why nanoelectronics?

The main advantages of scaling down electronic devices can be summarized as follows:

- *Increased speed of operation*—Reduction of the parasitic capacitances associated with nonconductive paths in an electronic device leads to a higher cutoff frequency. This enables a device to operate at much higher speeds.
- *Increased device density*—This is an obvious advantage that reduces size and cuts materials cost.
- *Reduced power dissipation*—It is due to lesser resistance in interconnects and currents flowing in smaller circuits.

However, there is a problem with scaling down silicon technology to nanometer sizes. Charge carriers in semiconductors are due to doping. Even if we would make use of highly doped material alone, only by chance could we find a *single* dopant in a 10-nm−sized transistor; another issue should be looked at is how to manage the heat generated from devices' density at a nanoscale level. Moreover, as a much lower voltage will be needed, it may result in difficulty in completely switching off these kind of devices (Box 7.2).

---

### Box 7.2 Advancing MOD devices

Nanotechnology is seen as a key in advancing eNose devices to a level that will match the olfactory systems developed by nature. Nanowire chemiresistors are seen as critical elements in the future miniaturization of eNoses. It is now also believed that single crystal nanowires are most stable sensing elements what will result in extending of life-time of sensors and therefore the recalibration cycle.

*Berger, 2007*

---

Naturally, we will be dealing with a very few numbers of charge carriers, if at all, and control of charge and electrical current on a single electron level will be required. Moreover, quantum phenomena will increasingly start to dominate the overall behavior of such structures. Tiny structures have a large surface-to-volume ratio, which is lethal for conventional semiconductor devices. Taking these facts, it is for certain that new approaches need to be developed.

Ari Aviram and Mark Ratner in the early 1970s started the idea of electronic circuit elements designed from *single molecules*. They explained the function of the molecular circuit in detail, and most likely, we can attribute the origin of the field of molecular electronics or at least molecular-scale electronics, to them (Aviram & Ratner, 1974) (Table 7.1).

The emergence of molecular electronics and spintronics is providing a challenge to traditional electronic manufacturing techniques. Significant reduction in size and the sheer enormity of numbers in manufacturing are the benefits of molecular electronics. Scientists predict that computers will be assembled using molecules, pushing technology far beyond the limits of silicon.

The statement made by Satish P. Nair in 2003 predicting the future of nanoelectronics is already occurring and progressing in this field, "The future of electronics is nanosized. Exciting nanofabrication techniques have unfolded different methods to engineer nanowires, quantum wells, and nanotubes which function as the building blocks of future nano-electronic devices. Molecular electronics can create devices that could be a thousand times smaller than current semiconductor-based devices. Molecular memories will also have a storage density million times that of today's best semiconductor chips." (Electronic newsletter "Space Daily," 2003). The progress with carbon nanotubes (CNTs)

**Table 7.1** Examples of nanoelectronics (Bhatia, Raman, & La, 2013)

| Technology | Size | Fabrication | Lithography | Problems |
|---|---|---|---|---|
| Carbon nanotube (CNT) | Few nanometers | Complex | Scanning tunneling microscope (STM) | Fabrication problem |
| Semiconducting nanowire (S-NW) | Few nanometers | Simple conventional technique | Conventional technique | Tunneling of current and subthreshold current |
| Quantum cellular automata (QCA) | Nanometers | Simple | Nanoimprint | Applicable in logic devices only |
| Molecular device | Nanometers | Complex | STM | Practical implementation |
| Graphene nanoribbon (GNR) | Few nanometers | Complex | STM | No problem |

*Source*: Edited from Bhatia, Raman, & La, 2013, The shift from microelectronics to nanoelectronics—a review. International Journal of Advanced Research in Computer and Communication Engineering 2, 11, November 2013. https://www.ijarcce.com/upload/2013/november/61-O-Inderdeep_Singh_-THE_SHIFT.pdf

and semiconductor nanowires has provided researchers with a model against which to gauge present and future nanoscale devices and systems.

Early breakthroughs in molecular electronics by industry giant Hewlett Packard and other major developers supported these predictions (Markhoff, 1999; Telecomeworldwire, 2001). Hewlett Packard has created a new kind of minute circuit for computer chips using nanotechnology. The company's research laboratory also developed the highest density electronically addressable memory.

Areas of nanoelectronics currently being studied include the fabrication of atomic wires; SET devices and atto-farad structures; and the study of spin-polarized electronics and magnetic nanostructures, all are likely to play an important part in the present and future electronic devices. By growing nanowires that are 20 to 200 nm in diameter (by comparison, a human hair is typically 50 to 100 μm thick), researchers are able to create the circuitry required for nanoelectronic devices.

Moore's law is still applicable today as it was decades ago in that the number of transistors are continuously doubling density per chip around every 2 years. This is due to the size of the transistor getting smaller with each generation. Not surprisingly, the claim made by Intel that 100.8 million transistors per square millimeter, as depicted in Fig. 7.1, illustrate the logic transistor density up to 2019. The prediction, as indicated by the graph, that the size of the transistor will continue to shrink and clearly will go below 10 nm as industries are developing fast into this direction.

Finally, research and development in nanoelectronics have been fueled by huge investments by various national governments, as is happening with nanotechnology in general. Countries in Europe and Asia, notably Japan and China, are allocating large sum of money in developing the field of nanoelectronics.

**Figure 7.1**
Logic transistor density. *Source: Redrawn and edited from Bohr, 2017, Moore's law leadership. Intel technology and amplified resists. https://newsroom. intel.com/newsroom/ wp-content/uploads/ sites/11/2017/09/ mark-bohr-on-intels-technology-leadership. pdf.*

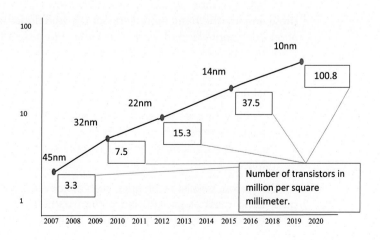

## 7.3 Lithography

Currently, optical lithography can be used for design rules down to near 100 nm and might even be used below this. This would imply an increasing process and mask complexity, increasing the cost of ownership. Alternative methods are actively being explored. The so-called "next-generation lithography" methods include extreme ultraviolet, X-ray, projection electron-beam and projection ion-beam lithographies, are just few examples (Box 7.3).

---

### Box 7.3 Today's optical lithography tools

Projection lithography tools come in two variations: step and repeat, and step and scan. In the step−and-repeat system (a stepper) the entire mask is illuminated and projected onto the wafer exposing one "die" (approximately 25 mm × 25 mm in size at the wafer). The light is then turned off and the wafer shifted (stepped) and the exposure process repeated. This cycle is continued until the entire wafer is exposed. In a step-and-scan system (a scanner) the imaging field size is reduced to a slit (typically on the order of 6 mm × 25 mm at the wafer), greatly facilitating the design and fabrication of the optical system. The mask and wafer stages are then scanned in opposite directions at the proper speeds such that the entire mask pattern is replicated in one scan, again creating an exposed die this time with a typical size of approximately 25 mm × 32 mm at the wafer. As with the stepper the light is then turned off and the wafer shifted over to an unexposed region where the die scan process is repeated.

*Naulleau, 2019*

---

There are a number of alternative routes to create smaller structures. Some of these employ the basic ingredients of the traditional technology: replica generation using some kind of mask. Currently, a number of technologies are actively working, which leave out the mask as an intermediate step.

For industrially applicable nanofabrication techniques, therefore the following minimal requirements are needed:

1. They must be able to produce millions or billions of these small structures in a quick, reliable, and cost-effective way.
2. These techniques must be able to connect these structures in a predefined manner.

For this reason, parallel techniques, such as conventional circuit lithography techniques, using mask alignment and pattern transfer seems to be a realistic way to achieve highly ICs. Serial techniques, such as scanning probe techniques or electron beam lithography, may be useful for mask making, or single-component fabrication, but do not supply adequate throughput for large-scale ICs.

## 7.4 Nanotubes

Nanotubes were discovered in 1991 by the Japanese electron microscopist Simio Iijima who was studying the material deposited on the cathode during the arc-evaporation synthesis of fullerenes. CNTs are fullerene-related structures which consist of graphene cylinders closed at either end with caps containing pentagonal rings. Examples of nanotubes are *single-layer nanotubes and nanotube "ropes" and nanohorns*. CNTs therefore are rolled-up sheets of graphite—that is, the same material that is used in pencils. A sheet of graphite is composed of carbon atoms arranged in a flat hexagonal pattern, similar to chicken wire mesh.

As has been already indicated, nanoelectronics has witnessed a shift toward molecular systems in recent years. Although the term molecular electronics is rather an old one, it is only recently that single molecules have become the focus of interest as nanoelectronics start to surface. This was triggered by research on CNTs. But before CNTs entered the scene, molecular electronics was the science of organic polymers, their synthesis, processing, and doping. With CNTs, we finally have a model system at hand that is equally of interest to chemists, material scientists, and physicists. However, CNTs are supramolecular objects for a chemist, and they are one-dimensional solids for a physicist. Currently and in the future, more of this supramolecular structure will be studied further at the single molecule level.

Theorists as well as well recent research have shown that nanotubes can be conducting or insulating, depending on their structure. This may lead to applications in nanoelectronics.

Wires (nanowires) are not possible for use in nanoelectronics, because they are susceptible to thinning and breakage. Despite recent interest in CNTs, they have variable electronic properties, depending on their orientation, reducing their functionality as electrical conductors.

One problem that plagues researchers looking to fashion circuit components from nanotubes is separating metallic tubes from the ones that are semiconducting. Common synthesis procedures produce spaghetti-like mixtures of nanotube ropes that are unusable for semiconductor applications because they contain both types of tubes.

Nanotubes can be metals or semiconductors, and because of their strong chemical bonds and satisfied valencies, the materials boast high thermal, mechanical, and chemical stability. In addition, CNTs can be efficient conductors, as a result of their nanodiameters, long lengths, and defect-free structures, which make them ideal one-dimensional systems (Box 7.4).

## Box 7.4 Carbon nanotube-based gas sensors

Upon interaction with molecules a charge transfer occurs between CNTs and molecules that can drastically alter CNTs' electrical conductivity. Gas sensors working with CNTs being in contact with two metal electrodes rely precisely on this phenomenon. In 2000 Kong et al. demonstrated, for instance, that a single semiconducting SWCNT increases or decreases its conductance when exposed to $NO_2$ or $NH_3$ gas, respectively. The sensor exhibits a fast response (down to below 1 minute) that can be attributed to the high surface area of the CNT. However, the most remarkable feature is probably the fact that it works at room temperature. Metal oxides have often been used for sensing $NO_2$ or $NH_3$; however, they need to work at temperatures above 200°C to achieve enough sensitivity. The sensor proposed by Kong et al. instead operates at room temperature with sensitivity as high as $10^3$, while it can recover in 1 hour upon annealing at 200°C or in 12 hours if left at room temperature under flow of pure argon. The high surface area, which grants CNT-based sensor a fast response, is probably also the cause of their slow recovery. In addition to annealing the sensor at high temperatures, another strategy to improve its recovery time is to illuminate it with ultraviolet radiation.

*(Camilli & Passacantando, 2018)*

Theoretical models have predicted that nanotubes could behave as ideal one-dimensional "quantum wires" with either semiconducting or metallic behaviors. Study of transmission electron micrograph images, however, has indicated that the nanotubes also incorporate kinks and defects into their walls.

Progress in nanotubes synthesis has now yielded single-walled nanotubes (SWNTs) with well-defined diameters, bringing the experimental situation much closer to that of the theoretical models. Measurements indicate that these materials do behave like one-dimensional wires. The SWNTs should also be more sensitive to defects, to the extent that defects may dominate the transport characteristics. According to Ghavamian, Rybachuk, and Očhsner, 2018:

*"Although both macroscopic and atomic scale defects reduce the homogeneity of CNTs and significantly alter their properties, mainly the mechanical properties, the addition of adatoms into the CNT structure is an effective way to tailor physicochemical properties of the tubes for specific applications such as field emission devices, n−p-type nanojunctions, electrical connectors, photo-optical sensors, and others. Doping of CNTs can considerably improve the chemical reactivity of CNTs which paves the way for their unique applications as miniature gas detector elements, effective catalytic materials, protein immobilizers, and others."*

In this work a scanning tunneling microscope (STM) tip was used as a sliding electrical contact to probe the length dependence of SWNT conductance. Although atomic defects were not directly imaged, sharp conductance transitions and heterojunction behaviors in the nanotube conductance are suggestive of the signatures of nanotube defects.

The following points are summary of some the characteristics of CNTs:

1. The CNT can capture and retain any gas molecules inside it.
2. The interlocking carbon-to-carbon covalent bonds make nanotubes one of the strongest known materials.
3. Adding CNT to other materials, such as plastic, can make them conductive to thermal and to electricity.
4. It can provide various useful functions in medicine and pharmacy such as for drugs and DNA delivery and tissue regeneration.
5. CNT are more conductive to electricity than copper materials.
6. Because of the covalent bonds, CNT have a very high melting point can be added to various materials to add strength to them.
7. It is possible to produce CNT on a bigger scale using methods such as:
   a. arc discharge and laser ablation;
   b. high-pressure carbon monoxide disproportionation; and
   c. chemical vapor deposition.
8. The diameter of CNT may range from 1 nm to several nanometers.
9. CNTs are very elastic and highly flexible

## 7.5 Quantum dots

Quantum dots are nanometer-scale "boxes" for selectively holding or releasing electrons. A quantum dot holds certain number of electrons. Since quantum dots are fabricated in solids, not in vacuum, of course, there are many electrons in them. However, almost all of these are tightly bound to atoms in the solid. The few electrons spoken of are extra ones beyond those that are tightly bound. These extra electrons could roam free in a solid where they are not confined in a quantum dot (Box 7.5).

Over the past 30 years, quantum dots have been transformed from laboratory curiosities to the building blocks for a future computer industry. Quantum dots are small metal or semiconductor boxes that hold a well-defined number of electrons. The number of electrons in a dot may be adjusted by changing the dot's electrostatic environment. Dots have been made ranging from 30 nm to 1 μm in size and holding from zero to hundreds of electrons.

Electronic devices, such as MOD, can be built using quantum dots. It has been observed in experiments and shown theoretically that reducing the dimensions of a quantum dot raises the effective operating temperature of the electron confinement device. Present day quantum dots are large enough

## Box 7.5 Introduction to quantum dots

Quantum dots (QDs) are ultrasmall-sized semiconductor nanocrystals made up of 100–10,000 atoms,[1] within the size range of 1.5–10 nm. QDs exhibit size unique optical properties due to changes in bandgap energy caused by quantum confinement effects. On absorption of light, electrons are promoted from the valence band (lower electronic energy state) to the conduction band (upper electronic energy state), producing an electron–hole pair, called an "exciton." When the electron and hole recombine, energy is released in the form of a photon (radiative recombination). In bulk materials the exciton can spread out over the delocalized lattice. However, when the particle size falls under the Bohr radius, the energy required to create an exciton increases. This effect is termed "quantum confinement," and it is typically observed in ultrasmall size, crystalline, semiconductor materials. Smaller QDs possess larger bandgap energy, thereby emitting photons of higher energy (blue shifted) and vice versa.

*(Maxwell et al., 2019)*

(approximately 1 to 10 μm long and wide) that they require cooling with liquid helium or, at least, liquid nitrogen, to cryogenic temperatures. However, for a practical technology with widespread applications based upon such quantum-effect devices, it will be necessary to achieve room temperature operation. This requirement implies that it is necessary to invent and manufacture molecular-scale quantum dots that are only approximately 1 to 10 nm in linear dimension. Such a quantum dot would probably be constructed as a single molecule, that is, a *molecular quantum dot*. Molecular quantum dots are one example of the next-generation technology known as Molecular-scale electronics. According to Reed and Tour (2000), chemical synthesis and molecular wires operate by allowing electrons to move nearly ballistically along the length of a chain of ring-like chemical structures with conjugated pi-orbitals.

Some researchers have suggested that it may be possible to insert chemical groups of lower conductance into such a molecular wire, creating paired barriers to electron migration through the chain. Such barriers might create a molecular quantum-effect device that would function in a fashion similar to solid-state resonance tunneling devices that already have been fabricated, tested, and used in prototype quantum-effect logic applications.

Work in the area of quantum-based devices for nanoscale metrology has already been directed to fabricating an ultrasmall superconducting quantum interference device for applications in a single-particle detection. The fabrication of such a device will be a significant achievement and should prove important in areas such as future nanoscale frequency standards, emerging quantum computer, and single-particle sensor technologies and in the study of adatom-surface interactions (*adatom is an atom located or adsorbed on a surface*).

# 7.6 Conductivity of molecules and single-electron transport

Another question to be asked is, "Can we manipulate single molecules so that their electronic capabilities can be tested and then applied?"

In nanostructures, electrical properties can be markedly different from their macroscopic equivalents, thereby revealing many novel effects. "Progress in the field has been hampered by two problems. The first has been in making robust, reproducible electrical connections to both ends of molecules. After this has been achieved, the next problem is knowing how many molecules there actually are between the electrical contacts." (Gust, 2001).

Various companies at the present are working on developing a quantum standard of current and capacitance using single electron transport (Box 7.6). In addition, research into these new technologies, which are likely to play an important part in the present and future electronic devices, includes the fabrication of atomic wires and the study of spin-polarized electronics and magnetic nanostructures. Other possible e-applications could range from quantum computing and so-called "secure" quantum communication, to devices for single-particle sensor technologies, nanoscale frequency standards, and the study of adatom-surface interactions. The quantum effects on which most of these devices are based are very weak, and the measurement technology is of paramount importance.

---

## Box 7.6 Single Electron

Single electron quantization effects are really nothing new. In his famous 1911 experiments, Millikan observed the effects of single electrons on the falling rate of oil drops. SET was first studied in solids in 1951 by Gorter and later by Giaever and Zeller in 1968 and Lambeand Jaklevic in 1969. These pioneering experiments investigated transport through thin films consisting of small grains. A detailed transport theory was developed by Kulik and Shekhter in 1975. Much of our present understanding of single electron charging effects was already developed in these early works. However, a drawback was the averaging effect over many grains and the limited control over device parameters. Rapid progress in device control was made in the mid-1980s when several

*(Continued)*

---

**Box 7.6 (Continued)**

groups began to fabricate small systems using nanolithography and thin-film processing. The new technological control, together with new theoretical predictions by Likharev and Mullen et al., boosted interest in single electronics and led to the discovery of many new transport phenomena. The first clear demonstration of controlled SET was performed by Fulton and Dolan in 1987 in an aluminum structure. They observed that the macroscopic current through the two-junction system was extremely sensitive to the charge on the gate capacitor. These are the so-called Coulomb oscillations. This work also demonstrated the usefulness of such a device as a single electrometer, that is, an electrometer capable of measuring single charges.

*(Kouwenhoven et al., 1997)*

## 7.7 The problem of contact

The results of various research related to the conductivity of molecules attached to wires is very well known—for example, DNA has been found to be everything from an insulator to a superconductor. Previous work to measure the electrical properties of small numbers of molecules has yielded a wide range of values for their conductivities. Most previous studies have relied on a "mechanical" contact between molecules and a metallic wire, where the two are simply pushed together. What is needed is a way to connect individual molecules on a molecular circuit board. However, by understanding the natural characteristics and behavior of molecules within a substance (Box 7.7), new methods in dealing with the

**Box 7.7 What are intermolecular attractions?**

Intermolecular versus intramolecular bonds

*Intermolecular* attractions are attractions between one molecule and a neighboring molecule. The forces of attraction which hold an individual molecule together (e.g., the covalent bonds) are known as *intramolecular* attractions. These two words are so confusingly similar that it is safer to abandon one of them and never use it. The term "intramolecular" would not be used again on this site.

All molecules experience intermolecular attractions, although in some cases, those attractions are very weak. Even in a gas like hydrogen, $H_2$, if you slow

*(Continued)*

> ## Box 7.7 (Continued)
>
> the molecules down by cooling the gas, the attractions are large enough for the molecules to stick together eventually to form a liquid and then a solid.
>
> In hydrogen's case the attractions are so weak that the molecules have to be cooled to 21 K ($-252°C$) before the attractions are enough to condense the hydrogen as a liquid. Helium's intermolecular attractions are even weaker—the molecules would not stick together to form a liquid until the temperature drops to 4 K ($-269°C$).
>
> (Clark, 2018)

structure are required, for example, for a molecular circuit board, and related design can be achieved successfully.

As early as 2001, in an article in the October issue of the journal Science, a team headed by Professor Gust reported a method for creating through-bond electrical contacts with single molecules and the achievement of reproducible measurements of the molecules' conductivity. This work started with a uniform atomic layer of gold atoms and attached long, octanethiol "insulator" molecules to it through chemical bonds, forming a fur-like coating of aligned molecules. They then removed a few of the insulators using a solvent and replaced them with molecules of 1,8-octanedithiol, a molecule that is similar but is capable of bonding with gold at both ends and acting as a molecular "wire." Tiny (two-nm) gold particles were then added to the solvent, where they bonded to the free ends of the 1,8-octanedithiol molecules, thus creating a bonded metallic "contact" at either end of the conducting molecules. A gold-coated conducting atomic force microscope probe—a conducting probe with an atom-sized tip—was then run across the surface, and conductivity was measured when it made contact with the gold particles. When electrical measurements were made on more than 4000 gold particles, virtually, all measurements fell into one of five groups (five distinct conductivity curves). The conductivity curves were distinct whole-number multiples of a single, "fundamental" curve. The fundamental curve represents conduction by a single molecule of octanedithiol attached to the two gold contacts. When more than a single molecule was bound, each additional molecule increased the current capacity by the single unit amount of current that could be carried by one molecule. When the probe encountered octanethiol "insulator" molecules, which could not bond with a gold particle, a much higher electrical resistance was recorded.

*"The experimental results closely agree with theoretical quantum mechanical calculations for the conductivity of these molecules, and this gives us confidence that current theories can provide useful guidance for future experiments," Gust noted. "The molecule becomes a much better*

*conductor when it is 'soldered' into the circuit by the bonds to gold at each end." "This suggests how we can wire single-molecule components into a molecular circuit board, and lays some important groundwork for doing practical molecular electronics." (Gust, 2001).*

Single supramolecular structures can be used to create switches and storage media. As has been shown already with DNA molecules, the trend toward using molecules includes biological macromolecules as well. The ability to manipulate and characterize single molecules is an important first step for the exploration of suitable molecular functions. A fully functional chip, however, requires the ability to assemble the molecules with high precision into a functional network.

## 7.8 Microelectronic and nanoelectronic devices

Hybrid microelectronic—nanoelectronic devices are the probable way that nanoelectronics should develop. For example, one innovative idea combines a solid-state quantum-effect nanoelectronic device, namely a resonance-tunneling device, with a conventional silicon bulk-effect transistor. This generates a multistate switching device, a "resonance tunneling transistor," that can be used to make circuits of much greater logic density than with conventional microelectronic transistor logic. The single electron transistor or SET is a type of switching device that uses controlled electron tunneling to amplify current. A SET is made from two tunnel junctions that share a common electrode. A tunnel junction consists of two pieces of metal separated by a very thin ($\sim$1 nm) insulator. The only way for electrons in one of the metal electrodes to travel to the other electrode is to tunnel through the insulator. Since tunneling is a discrete process, the electric charge that flows through the tunnel junction flows in multiples of the charge of a single electron.

Whenever electrons are constrained to a small region, the effects of energy quantization need to be taken into account. In other types of devices that work with far fewer electrons, energy quantization plays a much more important role. Quantum dots and resonant tunneling devices are, of course, two such devices.

## 7.9 Spintronics

Finally, there is a third direction in nanoelectronics, which is already receiving more attention at present, and it will continue to do so in the near future. This new field is called "spintronics." As indicated earlier, spintronics is concerned with electromagnetic effects in nanostructures and molecules caused by the quantized angular momentum (the spin) that is associated with all fundamental particles like, for example, the electron. The magnetic moment of a particle is directly proportional to its spin. Hence if we learn to manipulate not only charge but also spin on a single electron level, information may be stored and transported in the form of quantized units of magnetism (Box 7.8).

---

### Box 7.8 Spintronics

Spintronics is a relatively new field of electronics that has been rapidly developing over the past few years. To understand the potential of this new kind of electronics, it is useful to consider the basic properties of the electron. One of these properties is the charge, which has been exploited until now in conventional semiconductor electronics. The charge of a single electron is constant and local changes of charge can only happen via a current. However, the electron also has a second property, namely the spin. The spin can be seen as an angular momentum which—corresponding to a rotating charge—also induces a magnetic moment. Much in contrast to a macroscopic angular momentum the spin can only have two different (opposite) directions. It is quantized, with the axis of quantization given by an external magnetic field. So, whereas the electron charge has a constant value, the spin of an electron can have two different values called spin-up and spin-down, which can be "flipped" from one value to the other. This property allows for the local creation of a spin population without a current (in contrast to a charge), but it also means that a spin population can vanish over time, while a charge population in the absence of a current cannot.

*(Nile, 2016)*

---

# 7.10 Molecular diodes and molecular switches

The use of nanowires can revolutionize the way electronic devices function and maintained. The aim is to investigate the building of simple molecular diodes and molecular switches (MDMS) for the purpose of replacing electronic lithographic circuit. These kind of switches and diodes could be used for faster processing and energy saving, as well as cost reduction in various electronic and computing devices, in general and the proposed MOD design in this book, in particular.

## 7.10.1 Nanowire

Nanowires are discussed as part of a new nanoelectronic device, specifically for the construction of MDMS for the MOD, as mentioned earlier. In addition, this section discusses the structure and the type of element of nanowires and the working mechanism of MDMS (made from nanowires). Obviously, the selection of the most suitable nanowires to build the MDMS depend largely on the type of the device to be used for.

Basically, by connecting the last two molecules on either side of each nanowires, with current on and off, a simple switch or diode is ready to experiment with. One of the important tools needed is atomic force microscope (AFM and (electron tomography method) ETM), in addition to other equipment and tools, which will aid in constructing MDMS.

Current voltage, molecules, and molecules movement (plus other variation) play an important part in achieving current "flow" or "nonflow." Within the present electronic devices, the electric current pass between a positive and a negative point, that is, in ordinary conductive wire, the current face resistance along the way. A current passing through one string of single molecules has no resistant, as each electron pass individually and therefore no electron in the way to collide with, generating little or no heat at all.

In this way, energy will be saved, and therefore a very low voltage can perform the same task longer (if not better) in comparison with the average voltage required to run a normal switch or diode for a similar for purpose.

There are number of ways already well known in the making of nanowires. A brief look at some of these methods have been presented in this section of this chapter.

The molecular switches in our bodies are the key element for growth, body functions, and so on. Consequently, when these switches operate correctly, physical health is the norm. That means faulty switches on genes and elsewhere in the body are the common basis of many human disfunctions, which can be the causes of certain illnesses.

Learning from nature of how molecules arranged themselves and work as "switch/diodes" to start certain function, will aid us tremendously in a research such as this one. If we can learn more and understand how molecular diode/switch work in nature and apply what we learn to the types of diodes and switches we need to design, then our task in this field would be much easier. After all, whether an electronic or biological switch in the nanoworld, both use molecules as a switching mechanism for different purpose.

Molecule switching if used on computing system can also provide the basis of logic and memory, as we can trap electron using molecules for a longer period. For this reason, designing molecular switches and diodes will bring us another step closer to molecules computing, an industrial revolution which will bring great benefit.

New techniques in synthesizing and manipulating certain types of molecules, in which electric current flow much more easily from one end of a molecular to the other, than it does in reverse, the asymmetry mean that the molecule act as rectifier and could be used as an electronic component, that is, a diode (Box 7.9).

## Box 7.9 Nanowire properties

Depending on what it is made from, a nanowire can have the properties of an insulator, a semiconductor, or a metal. Insulators would not carry an electric charge, while metals carry electric charges very well. Semiconductors fall between the two, carrying a charge under the right conditions. By arranging semiconductor wires in the proper configuration, engineers can create transistors, which either acts as a *switch* or an *amplifier*.

Some interesting and counterintuitive properties nanowires possess are due to the small scale. When you work with objects that are at the nanoscale or smaller, you begin to enter the realm of quantum mechanics. Quantum mechanics can be confusing even to experts in the field, and very often, it defies classical physics (also known as Newtonian physics).

*Strickland (2007)*

Using a highly stabilized microscope, we can learn how to rotate one conducting molecule on each end of the single nanowire so that the two can have the same opposing sides at the end of each nanowire, that is, two positive or two negatives. The reason for this is because similar sides of most molecules repulse each other and therefore disturbing the flow of the electron orbital field, that is, provide the switching off function we want.

Clearly, if we are experimenting with nonconductive molecules, then we may use gold particles (which are much far bigger than the molecules), which will allow the current to flow.

### 7.10.1.1 Making nanowire (brief description)

Nanowires can be made ready for the next step for diodes and switches.

As mentioned previously, there are many ways of making nanowire; however, the example described below has been chosen for the simple process used in making them.

1. Using nanoparticle of cadmium telluride (CdTe) that spontaneously self-assembled into crystalline nanowires and using CdTe nanocrystals stabilized by thioglycolic acid ($C_2H_4O_2S$). By removing destabilizer and leaving the crystal at a room temperature in darkness, for a period of time, the nanoparticle rock reorganized into set of chain and crystallized into nanowires.
2. Making nanowire composed of two different semiconductors, silicon and silicon germanium.

3. Using a very small 2.5-cm tube furnace, nanowire grow using a hybrid pulse laser ablation/chemical vapor-deposition process.

Putting inside the furnace a silicon wafer coated with a thin layer of gold then raising the temperature inside the furnace, the gold coating melt and form an alloy with the silicon. Still the alloy in a liquid state, it will start to break up into nanometer-sized droplets. The two-semiconductor vapor will then be settled and condense deposit around the gold droplet. The result of this is nanowire segments. If the laser is off, only one element is deposited, that is silicon on the gold particle; however, if the laser is on, the deposit is both elements, that is, silicon and germanium. This second method produce thicker nanowires (more layer molecules than the first method used).

To make sure a proper and good conducting take place, a one layer (or more) of nonconductive molecules can be used to seal nanowire from others layer of conductive material, including neighboring another nanowire.

## 7.10.2 Molecular diodes and switches

### 7.10.2.1 Basic molecular diodes

One piece of nanowire can be made to behave as a diode as the current pass through it. The process works in the following manner, in certain molecules for example, nitrogen and carbon:

If there is a current going through in one direction (switch on), then as we reverse the current, the molecules change position (rotate) and the electron field (internal energetic line-up of orbitals) of the molecules will change, and consequently, the flow of the current will slow down and/or stop completely (switch off). This kind of conductivity, that is, one direction only, is what a normal diode is designed to function in any electronic device. To increase conductivity, gold particles can be used (or different type of molecules coated with gold instead) as part of the nanowire.

Fig. 7.2 illustrates how a current can pass easily from left to right, while Fig. 7.3 shows current has been reversed, and with it the current stopped flowing.

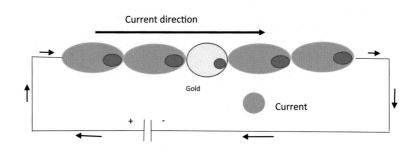

**Figure 7.2** Schematic diagram of nanowire (diode) with current going through (*gray*).

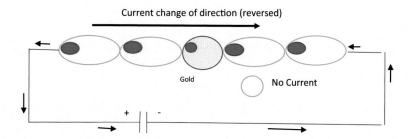

**Figure 7.3**
Schematic diagram of nanowire (diode) with no current going through (*white*).

### 7.10.2.2 Basic molecular switches

We can use a similar approach with the switches, that is, how molecular diode can be changed to work as a molecular switch, but with some improvement. This means we can make the current flow in two different directions, with the capability of being turned off and on, as required.

We already know that by changing the voltage—as the current going through a nanowire—a change in the conductivity will take place. The changes of voltage will change the position of molecules and, consequently, the internal energetic line-up of orbital, that is, switch off. This is very much like what happens when we change the current direction in a molecular diode.

Basically, we already have a simple molecular switch simply by varying the voltage on each side of the nanowire, that is, the flow of electron, which the molecule resists, depend on the voltage applied to them. Varying the voltage, as mentioned earlier on, it is possible to repeatedly change the molecules anytime from a conducting to a nonconducting level, like the function of any electrical switch.

If there is a problem needs to be looked at, then it would be the voltage variation and how to control it. This is not the main part of discussion but rather a secondary issue, which can be addressed during the test.

One solution for this problem is to add a third and a fourth nanowire. These two nanowires would be made up from molecules with a slightly different conductivity.

Even though the two nanowires are receiving current from the same source, they will have some differences in the amount of voltage. This is because there is no resistance within the majority of nanowires. The flow of electron will be almost the same. The infinitesimal difference in timing and voltage will complete the basic requirement for a simple molecular switch which function. This means there is a fast change in voltage from different conductivity, which will give the switch-off state, and at the same time the voltage would be back to the required level for conductivity in a fraction part of a nanosecond (Fig. 7.4). (Box 7.10).

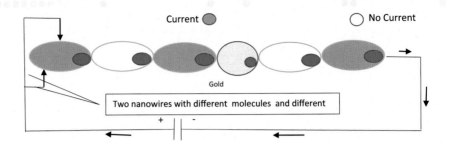

**Figure 7.4**
Schematic diagram of nanowire with a continuous change in voltage as the state of "on" and "off" can be induced.

---

**Box 7.10 Chemically activated switches**

Biology uses chemically responsive switches and machines to regulate a myriad of processes, including the creation of proton gradients, protein folding, and viral entry and translocation. These biological processes offer excellent temporal control via well-regulated signaling cascades as well as fast and efficient switching sequences. In principle, artificial chemically activated switches can be used to mimic all these processes and expand their use in nonbiological areas. The bistability of these systems, their complete and reversible switching based on a myriad of benign and biocompatible inputs, which when selective allow for sensing, fatigue resistance, and use in inter-system communication, are added benefits to their light-activated counterparts. One major drawback of such chemically activated systems, however, is the accumulation of sacrificial species in solution, a problem nature has solved through sequestration and excretion. Accumulation of waste products tends to reduce the efficiency of chemically activated switches over many cycles. If we, as chemists, wish to mimic and eventually compete with biology, then we need to address this issue, in addition to developing new chemically activated switches capable of complex functions. Here, we survey new architectures of chemically activated molecular switches and the properties (e.g., switching efficiency, thermal stability, and switching cycles) needed to achieve these goals.

*(Harris, Moran, & Aprahamian, 2018)*

---

## 7.11 MOD and nanoelectronic system

For the miniaturization of MODs to be part of any device that may require the detection of harmful or nonharmful airborne substance, even on a very small scale, then nanoelectronic can be the answer in designing such a system. The approach is somehow similar to the way that olfactory systems of mammals and insects work. This may mean using nanotubes or nanowires (metal oxide) as the basis for designing the sensors. The general principles will be microprocessing; however, the need for advanced electronic microscopy is an important part of the above-mentioned approach. The following section provides a brief detail concerning one of these important tools, that is, the electronic microscope.

## 7.11.1 Electron microscope

The idea of producing a microscope able to image atoms was conceived in the first part of the 20th century. Ernst Ruska (1906−1987) had shown in his PhD thesis the potential for electrons to be used in a microscope. In 1931 Max Knoll and Ernst Ruska constructed the first electron microscope. During 1933 Ruska built the first electron microscope with and accelerating voltage of 75 kV it exceeded the resolution of a light microscope.

The electron microscope established itself as a powerful instrument in the 1950s as ultramicrotomes (1951) and first diamond knife (1954) were built.

The first electron microscope which could view biological samples was made in 1934, followed by the first scanning microscope.

The year 1982 witnessed the invention of the scanning probe microscope (SM), followed in 1986 by the invention of the AFM. For the working principle of electron microscope, see Box 7.11.

---

### Box 7.11 Working principle of electron microscope

Electron microscopes use signals arising from the interaction of an electron beam with the sample to obtain information about structure, morphology, and composition.

1. The electron gun generates electrons.
2. Two sets of condenser lenses focus the electron beam on the specimen and then into a thin tight beam.
3. To move electrons down the column an accelerating voltage (mostly between 100 kV and 1000 kV) is applied between tungsten filament and anode.
4. The specimen to be examined is made extremely thin, at least 200 times thinner than those used in the optical microscope. Ultrathin sections of 20−100 nm are cut, which is already placed on the specimen holder.
5. The electronic beam passes through the specimen, and electrons are scattered depending upon the thickness or refractive index of different parts of the specimen.

*(Continued)*

---

> **Box 7.11 (Continued)**
>
> **6.** The denser regions in the specimen scatter more electrons and therefore appear darker in the image since fewer electrons strike that area of the screen. In contrast, transparent regions are brighter.
> **7.** The electron beam coming out of the specimen passes to the objective lens, which has high power and forms the intermediate magnified image.
> **8.** The ocular lenses then produce the final further magnified image.
>
> *(Aryal, 2018)*

### 7.11.1.1 Scanning probe microscope

How do the STM and AFM work? Well, when we bring a probe (tip) very close to a surface (near 1 nm), different physical phenomena occur. This may be generation of a small electric current between the probe and the surface (in STM) or else a small force between the tip and the surface (in AFM). If we move the tip over a surface very precisely, and for example, keep the current generated (in STM) or the force (in AFM) constant, we can measure the in-and-out movement of the tip very carefully, and this allows us to obtain information at the nanometric level. This is why the name "scanning probe" is used.

If we polarize a conducting or semiconducting sample with a conducting tip, then electrons are transferred between the surface and the tip. This current is produced by a quantum phenomenon known as the "tunnel effect." When scanning the surface, a piezoelectric ceramic attached to the tip allows very small displacements to be made and accurately measured; so we can record the tunneling current variations, which are related to the tip-surface distance (or, alternatively, we can keep the current constant and measure the displacements). As noted earlier, this technique is called STM. In the case of very planar surfaces, it is possible to actually see atoms and even to manipulate them.

The various positional devices used in SPMs are especially designed to be stiff, to image individual atoms, despite the problems of thermal noise which exist in this kind of environment.

The recorded images are digital, generally 512- $\times$ 512-point plots. Each point can be given a gray-scale value between 0 and 255 so that the greater the value the more elevated is the point and vice versa. Therefore, in this way, we can represent 3D information in 2D. When we present a 2D image, we also include the Z range to determine the height difference between black and white areas. Alternatively (and more commonly nowadays), color coding of height variations can be used.

AFM uses the same tools of tip and surface. In this case a tip attached to a cantilever approaches the surface. An interaction involving different types of forces (van der Waals force, capillary force, friction, etc.) appears between them. By pushing on the structure being scanned and feeling how hard it pushes back, the AFM can build up an image of stiff structures. In measuring the force variation during scanning, we record the height variations on the surface.

Four things to note about AFM are as follows:

1. It does not work very well when the applied force is strong enough to deform the structure.
2. It is very useful if the precise molecular structure of the tip can be controlled, so that the precise nature of the tip surface interaction can be well defined.
3. This technique has the advantage of being able to function under widely varying (ambient, liquid, and electrochemical) environments and on either conducting or insulating samples.

With an AFM an approach for positioning molecular building blocks (in vacuum or in a liquid medium) can be achieved. This is because, as the AFM touches the surface of the structure being probed, it can also change that surface. With today's SPM capability, scientists have already built nanoscale structures. Examples of this kind of structures are the near—field scanning optical microscope, scanning thermal microscope, scanning capacitance microscope, magnetic force and resonance microscopes, and the scanning electrochemical microscope.

SPMs can "see" the structure that is being built and provide constant feedback to the operator, which help them to detect and correct errors. We are still far behind when it comes to the actual work and speed of SPMs, as their work should be done in a fraction of a second, rather than what is happening in most cases today (as it may take a number of hours to arrange a few molecules or atoms).

Whether through self-assembly, by improvements in SPMs, some hybrid approach, or perhaps by some other path, we are moving from an era of expensive and imprecise products to an era of inexpensive products of molecular precision.

## References

Aryal S. (2018). Electron microscope—definition, principle, types, uses, images. <https://microbenotes.com/electron-microscope-principle-types-components-applications-advantages-limitations/>.

Aviram, A., & Ratner, M. (1974). Molecular rectifiers. *Chemical Physics Letters*, 29(2), 277—283, ISSN 0009—2614. Available from https://doi.org/10.1016/0009-2614(74)85031-; <http://www.sciencedirect.com/science/article/pii/0009261474850311>.

Berger, M. (2007). Nanotechnology electronic noses. *Nanoweek*. <https://www.nanowerk.com/spotlight/spotid = 3331.php>.

Bhatia, I., Raman, A., & La, N. (2013). The shift from microelectronics to nanoelectronics—a review. *International Journal of Advanced Research in Computer and Communication Engineering, 2*(11). November 2013 <https://www.ijarcce.com/upload/2013/november/61-O-Inderdeep_Singh_-THE_SHIFT.pdf>.

Bohr, M. (2017). Moore's law leadership. *Intel technology and amplified resists.* <https://newsroom.intel.com/newsroom/wp-content/uploads/sites/11/2017/09/mark-bohr-on-intels-technology-leadership.pdf>.

Camilli, L., & Passacantando, M. (2018). Advances on sensors based on carbon nanotubes. *Chemosensors, 6*(4). Available from https://doi.org/10.3390/chemosensors6040062. <https://www.mdpi.com/2227-9040/6/4/62/htm> MDPI.

Clark J. (2018) Intermolecular bonding—van Der Waals forces. <https://www.chemguide.co.uk/atoms/bonding/vdw.html>.

Electronic newsletter "Space Daily". 4 November 2003. <http://www.spacedaily.com/news/nanotech-03b.html>.

Ghavamian, A., Rybachuk, M., & Oöhsner, A. (2018). Defects in carbon nanotubes. In Jan Stehr, Irina Buyanova, & Weimin Chen (Eds.), *Woodhead publishing series in electronic and optical materials, defects in advanced electronic materials and novel low dimensional structures* (2018, pp. 87–136). Woodhead Publishing. <https://reader.elsevier.com/reader/sd/pii/B9780081020531000041?token = B717B54D55E72C6D59C6EF76453E7C2B082C5342CB890E070EB4919B9B5E98F178D139D03C02B916611D7FCEF88B3489>.

Gust (2001). Electrical conductivity of single-molecule 'wire' accurately measured using new techniques. Press release from Arizona State University. 18 October 2001. <http://clasdean.la.asu.edu/news/molecular.htm>.

Harris, J., Moran, M., & Aprahamian, I. (2018). New molecular switch architectures. *Proceedings of the National Academy of Sciences of the United States of America, 115*(38), 9414–9422. first published July 16, 2018 <https://www.pnas.org/content/115/38/9414>.

Kouwenhoven, L., Marcus, C., Mceuen, P., Tarucha, S., Westervelt, R., & Wingreen, N. (1997). Electron transport in quantum dots. *Advanced Study Institute on Mesoscopic Electron Transport.* <https://courses.physics.illinois.edu/phys598MN/sp2015/readings/Dots_Review.pdf>.

Markhoff J. (1999). Researchers are crazy about molecular electronics (Digest of New York Times article by J. Markhoff, 1 November 1999), Digest prepared by J. S. Davis. CALMEC Scientific Forum. <http://www.calmec.com/sfarticle05.htm>.

Maxwell, T., Campos, M., Smith, S., Doomra, M., Thwin, Z., & Santra, S. (2019). Quantum dots. *Micro and Nano Technologies, 2020*, 243–265. <https://www.sciencedirect.com/science/article/pii/B9780128166628000151>; <https://www.sciencedirect.com/science/article/pii/B9780128035818020555>.

Naulleau, P. (2019). Optical lithography. *Comprehensive Nanoscience and Nanotechnology, 2* (2019), 387–398. (Second Edition) <https://www.sciencedirect.com/science/article/pii/B9780128035818104333>.

Nile, T. (2016). Spintronics in semiconductor nanostructures. *Reference Module in Materials Science and Materials Engineering.* <https://www.sciencedirect.com/science/article/pii/B9780128035818020555>.

Reed, M., & Tour, J. (2000). Computing with molecules. *Scientific American, 282*(2000), 86. <http://www.sciam.com/article.cfm?articleID = 0000B2EC-6247-1C74-9B81809EC588EF21&pageNumber = 1&catID = 2>.

Strickland. (2007). How nanowires work. *HowStuffWorks.* <https://science.howstuffworks.com/nanowire.htm>.

Telecomeworldwire. (2001). Brief article in electronic newsletter—18 July 2001. <http://www.findarticles.com/cf_dls/m0ECZ/2001_July_18/76606945/p1/article.jhtml>.

# Machine olfaction device nanostructure coating

It has become apparent that nanostructured ceramic and metallic coatings offer lasting protection and consequently longer life cycle for machine olfaction devices (MODs), especially when used within a harsh environment. For this reason, there is a growing interest across the world in developing new types of nanostructured coating with high hardness and high toughness. Nanostructured coatings have a large range of potential applications, which can be used within variety of fields such as construction, architecture, and devices, with an exterior protection including photocatalytic nanocoatings, self-cleaning nanocoatings, UV-protection nanocoatings, antigraffiti nanocoatings, superhydrophilic and hydrophobic nanocoatings, and antireflection nanocoatings; however, this chapter of the book is focused on coatings for the MOD device to be located in a harsh working environment.

Continuous improvement in the properties of such coatings has been steadily taking place but especially so within the past three decades. Many different systems have been developed. Of particular interest regarding this work are ceramic/metal nanocomposite systems, such as MeN/Me (molybdenum–nitrogen) coatings.

This chapter is concerned with molybdenum nitride ($MoN_x$) coatings with relatively high nitrogen content. It is part of further, wider work to develop an understanding of the influence of nanocomposite coating composition and structure on the hardness, modulus, and resulting tribological behavior of the coated materials. The specific aim was to characterize a number of coatings, sputter-deposited with different compositions. Characterization was undertaken using a combination of X-ray photoelectron spectroscopy (XPS), scanning electron microscopy (SEM), and transmission electron microscopy (TEM) (Box 8.1).

## 8.1 Molybdenum nitride

The $MoN_x$-based nanocomposite films may have several structures, depending on the amount of nitrogen present. With increasing amounts of nitrogen, they may

1. contain no nitride phase (i.e., the nitrogen is held mainly in interstitial solid solution in molybdenum);

**Introduction to Machine Olfaction Devices.**
**DOI:** https://doi.org/10.1016/B978-0-12-822420-5.00008-8

---

## Box 8.1 Advantages and disadvantages of nanocoating.

*Main advantage*

- Better surface appearance,
- good chemical resistance,
- decrease in permeability to corrosive environment and hence better corrosion properties,
- optical clarity,
- increase in modulus and thermal stability,
- easy to clean surface,
- antiskid, antifogging, antifouling, and antigraffiti properties,
- better thermal and electrical conductivity,
- better retention of gloss and other mechanical properties like scratch resistance,
- antireflective in nature,
- chromate and lead free, and
- good adherence on different type of materials.

*Disadvantages*

- Main problem in using nanoparticles for coating purpose is dispersion and stability of nanoparticles. Agglomeration may take place because of high surface energy possessed by nanoparticles because of their large surface area.
- Pigments may lose their color on reducing their size to nanolevel and hence will lose their opacity.
- Stable binder is required to inhibit photocatalytic activities of nano-$TiO_2$.
- Hardening problems of ultrafine powder.
- Extensive use of nanoparticles may result in new type of environmental problems, such as newer type of toxic materials and other environmental hazards. Ultrafine particles can catalyze chemical reactions inside the body, which might be dangerous (Chaudhari, Patil, Raichurkar, & Daberao, 2018).

---

2. contain small amounts of lower nitrides (i.e., $Mo_2N$); and
3. contain mainly one nitride phase with a small amount of molybdenum or of a second nitride phase.

For nitrogen concentrations from 0−33 at.% under equilibrium conditions, the Mo−N phase diagram predicts there to be a two-phase composition of Mo plus gamma-molybdenum nitride ($\gamma$-$Mo_2N$). However, previous work has shown that for coatings deposited by PVD methods, such as sputtering, formation of the nitride phase can be suppressed at low N contents.

At higher nitrogen contents (i.e., >15 at.%), $MoN_x$ coatings may be expected to have a structure consisting of $\gamma\text{-}Mo_2N$ and Mo nanocrystallites. Because the coatings under investigation are known to have a relatively high nitrogen content, the aim was to investigate the nanostructure formed (Box 8.2).

---

## Box 8.2 ß-Molybdenum nitride

Transition-metal nitrides in general, and Mo nitride in particular, exhibit a combination of properties that have resulted in multiple applications such as coatings/structural components, high performance magnets, in electronic and optical devices, and as catalytic materials. Conventional preparative routes involved the following:(1) high temperature (1400K−1900K) reaction of the base metal and elemental nitrogen; (2) carbothermal nitridation of metal oxides; or (3) a self-propagating high temperature synthesis. Alternative methods that can operate under milder reaction conditions have drawn on controlled temperature programmed procedures.

A combination of $NH_3 + H_2$ has been the most widely employed reacting gas in reduction−nitridation processes. However, the use of a $N_2 + H_2$ mixture circumvents the heat transfer problems associated with the endothermic $NH_3$ decomposition. The methodologies applied to date have generated a combination of $Mo_xN_y$ phases, principally metastable cubic ($\gamma\text{-}Mo_2N$)and hexagonal (d-MoN) structures. Although $\gamma\text{-}Mo_2N$ is the most commonly synthesized form by thermal treatment of $MoO_3$, there is also evidence in the literature for the formation of a body-centered tetragonal b-nitride phase with a Mo/N ratio in the range 2.0−2.6 (Cárdenas-Lizana, Gómez-Quero, Perret, Kiwi-Minsker, & Keane, 2010).

---

## 8.2 Analytical techniques and samples

All samples should be prepared, examined, and analyzed using three techniques, namely TEM, SEM, and XPS. After selecting three $MoN_x$ samples, these should be deposited onto silicon substrates by sputtering with varying chemical compositions. The following sections will provide background information about the techniques and TEM specimen preparation.

### 8.2.1 Transmission electron microscopy

The most important tool used to analyze these samples is the TEM.

A brief general summary of the ways in which TEM can be used as an analytical tool will be provided, as well as discussing two aspects of TEM, particularly relevant to this chapter, namely the use of electron diffraction patterns (and associated theory) and the preparation of TEM samples.

### 8.2.1.1 TEM as an analytical tool

The uses of TEM as an analytical tool are well described in standard textbooks and on the websites of manufacturers and the brief summary here draws on these sources.

In TEM a beam of electrons passes through a thin sample of material. Most electrons do not interact with the material, lose no energy and pass through the sample unaltered, but others undergo one of several possible interactions.

The electrons that are not scattered are used to form a *bright-field* image. For a crystalline sample, some electrons will satisfy the Bragg equation and will be diffracted as a result. These electrons can be used to form an electron diffraction pattern or a *dark-field* image.

1. Using the TEM, we can implement various imaging and analytical techniques, namely: bright field images, dark field images, high angle annular dark field images (Z contrast), Energy dispersive spectroscopy (EDS) analysis, electron energy loss spectroscopy (EELS), EELS imaging, EDS mapping, high-resolution electron microscopy structural images, electron holograms, selected area diffraction, nanodiffraction, convergent beam diffraction, and high dispersion diffraction.
2. There are various sample preparation techniques, namely: low angle ion beam milling, dimpling, plasma cleaning, disk cutting, mechanical polishing, electropolishing, and focused ion beam milling.
3. Image processing and analysis techniques include high-resolution image simulation, diffraction pattern simulation, crystal modeling, contrast improvement and noise reduction, particle and area counting, convergent beam and Kikuchi pattern simulation.

In regard to this part of the book, the TEM instrument should be used to form selected area diffraction patterns, bright-field images, and dark-field and High-resolution transmission electron microscopy images.

### 8.2.1.2 Electron diffraction patterns

The electron diffraction pattern from a nanocrystalline sample consists of a series of rings (in places broken up into "spots") from which we can determine the phase composition. The angle through which the electrons are diffracted gives rise to the radius $R$ of a particular ring or spot formed (i.e., $R$ is the distance between the central spot and the diffraction ring or spot). Using formulas discussed further, we can use the $R$ value to determine the lattice-plane spacing ($d$ value) for a particular diffraction spot or ring.

The electron diffraction pattern provides information on the microstructure of the specimen being examined. However, there is a difference between amorphous and nonamorphous (crystalline) materials. In the first case, the arrangements of atoms (or molecules) are arranged entirely randomly; in the second case, there is a distinct repeating structure of atoms.

From diffraction patterns, we can in principle:

1. measure the average spacing between layers of atoms;
2. determine the orientation of the crystals;
3. find the crystal structure of an unknown material; and
4. measure the size, shape and internal stress of small crystalline regions.

## 8.2.2 Electron diffraction theory

The theories of electron diffraction and X-ray diffraction (XRD) are similar. The following provide a brief description:

In a crystal, the atoms are arranged in a regular fashion. The atomic positions relate to the positions of a regular array of points in space. The array is called a *lattice*, and the points are called *lattice points*. Sets of parallel mathematical planes (called *lattice planes*) may be drawn through the lattice points in such a fashion that each lattice point is on a lattice plane. This may be done in many different ways. Each possible set of lattice planes can be labeled by a set of three integers $(h,k,l)$ called *Miller indices*. For each possible set of lattice planes, there is a characteristic distance between adjacent planes called the *lattice plane spacing* and denoted by $d$ (or by $d_{hkl}$ when it is necessary to be more precise).

There are 13 types of regular crystal lattices. Of most interest in this chapter is the *cubic* lattice. We can think of this as a three-dimensional stack of many small cubes; there are lattice points at each corner of every cube. The cube has a side of length equal to the *lattice constant a*. For a cubic lattice, it can be shown that the $(h,k,l)$ set of lattice planes has a lattice-plane spacing given by:

$$d_{hkl} = a/\left(h^2 + k^2 + l^2\right)^{1/2}. \tag{8.1}$$

It follows that:

$$\left(a/d_{hkl}\right)^2 = h^2 + k^2 + l^2. \tag{8.2}$$

When an electron beam or X-ray beam, of wavelength $\lambda$, impinges onto a crystal, most of it goes straight through, but "planes" of the lattice diffract some electrons or x-rays. Each set of indexed lattice planes in the crystallite diffracts part of the beam by a specific angle, as shown in Fig. 8.1

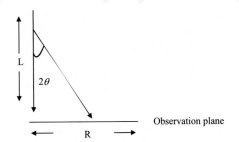

**Figure 8.1**
Part of the electron beam in TEM is diffracted by a set of indexed lattice planes in the crystallite.

It has been shown by Bragg that the *angle of deflection* ($2\theta_{hkl}$) is related to the lattice plane spacing by:

$$\lambda = 2d_{hkl}\sin\theta_{hkl}. \qquad (8.3)$$

Different lattice planes give rise to different *intensities* of the diffracted beam if the crystallites are randomly oriented; the reflections from planes corresponding to a given Miller index form a spotty ring of radius $R$ in the *diffraction pattern* formed in the *observation plane* in Fig. 6.2. The geometry of the situation, as shown in Fig. 6.2, means that $R$ is given by

$$R = L\tan(2\theta_{hkl}), \qquad (8.4)$$

where $L$ is called the *camera length*.

Combining Eqs. (8.3) and (8.4) and using the approximations (valid for small $\alpha$) $\tan\alpha = \alpha$, $\sin\alpha \approx \alpha$, we obtain the formula:

$$d_{hkl} = (Ll)/R_{hkl} = C/R_{hkl}. \qquad (8.5)$$

The quantity $C$ is called the camera constant and is characteristic of a particular experimental configuration. $C$ is the same for all rings in a given observed diffraction pattern. For a given experimental configuration, $C$ can be determined by a sample of known composition. However, with a TEM, the value of $C$ can vary slightly, depending on experimental parameters. Small calibration errors can exist if $C$ is not determined for the experimental conditions used to obtain the diffraction patterns.

Databases exist, particularly for XRD that gives the following:

1. information concerning the structure of the material;
2. the Miller indices of the reflections that may be observed;
3. the values of the corresponding lattice-plane spacing; and
4. the relative intensities of the various reflections.

For electron diffraction, the relative strengths will in general be different, but the other data are valid (Box 8.3).

The easiest way to identify the diffraction pattern of an unknown material is often to compare the experimentally observed *d* values with the tabulated *d*

> ## Box 8.3 Electron diffraction and scattering.
>
> The great advantage of electron diffraction over using X-rays and neutrons is that in the latest generation of electron microscopes, electron beams can be focused to a spot of only 1 Å in diameter. Hence local information about the specimen can be obtained with atomic resolution. The combination of electron diffraction, imaging, and electron energy loss spectroscopy enables us to answer the key questions of "where are the atoms?" and "what are the atoms?" for any material, Electron scattering is a key technique for not only basic science but also applied science: as electronic and other devices shrink to nanoscale dimensions, electron scattering is now an essential tool for analyzing the local structure of such devices. There can be little doubt that as the importance of nanoscience and nanotechnology increases, the already widespread use of electron scattering and diffraction will increase still further. It will be an even more important technique in the future than it has been in the past (Humphreys, 2002).

values for all different structures; that might conceivably be present in the observed specimen, taking account of expected relative ring strengths (but also bearing in mind the possible existence of a systematic calibration error that may be as high as 10% or more).

## 8.3 TEM sample preparation

To be able to examine any of the abovementioned samples, a process of thinning each sample to less than 50 μm has to be done via the following stages:

1. A small piece is cut from the wafer of the coated silicon sample. The thickness of the sample is measured using a micrometer. A very small piece of wax was then used to mount the sample piece on a pallet.
2. The sample piece is polished in four stages. The first stage is to use grade 1200 silicon carbide (SiC) paper. This is followed by polishing with grade 2400 and then grade 4000 paper. After each stage of polishing an ultrasonic bath was used to remove any deposit of material on the surface of the sample and a micrometer used to measure the thickness of the polished sample. If it is less than 50 μm, then the sample underwent a final diamond polish. If not, then SiC paper is employed again to reduce the thickness further.
3. Using a tiny amount of glue the sample is mounted on a small copper disk, containing a hole in the middle, and then left to dry in a warm place for a minimum of one hour. After the glue has dried completely, the sample is

put in acetone for final cleaning, before being analyzed by TEM. Obviously, the sample may need to be examined (during the various previous stages) to check for the hole itself on the sample and to make sure that the sample is clean and has no wax deposit. This is done using a basic light microscope.

4. Finally, the sample is cleaned once more using the ultrasonic bath. On some occasions, it may be necessary to leave it for a while in acetone for additional cleansing, before putting the sample in the precision ion polishing system to create a hole in the sample by ion beam bombardment. Two beams were used, one above and one below the specimen, operating with an acceleration voltage of 5 keV and a current of 1 mA. The machine bombards the sample with an argon beam (for about 2 hours, for a thin sample) until a hole in the sample has been created.

5. The sample is now sufficiently thin to be examined by TEM.

## 8.4 Scanning electron microscopy

SEM is used for the imaging and analysis of small features on solid surfaces. Sample preparation is relatively easy, as most SEM samples only need to be conductive. The combination of higher magnification, larger depth of focus, greater resolution, and ease of sample observation makes the SEM one of the most widely used instruments in research today.

SEM provides magnifications up to $\times 300,000$ and has a high spatial resolution of less 2 nm (as measured by gold particles on carbon). Compositional analysis of a material may also be obtained by monitoring X-rays produced by electron–specimen interactions. Characterization of fine particulate matter in terms of size, shape, and distribution, as well as statistical analyses of these parameters, maybe performed.

## 8.5 X-ray photoelectron spectroscopy

XPS involves the bombardment of a sample surface with X-rays and measurement of the intensity of photoelectrons and Auger electrons emitted from that surface as a function of their energy. As a technique, XPS can in principle provide not just elemental analysis but also chemical information. Thus, it can determine chemical states, for example, it can distinguish Si-Si bonding from Si-O or Si-C bonding.

XPS is used to provide information about the elemental composition of the coatings. The XPS spectrometer is used to obtain a survey spectrum, from which the elements present can be determined. Individual spectral peaks are then examined at higher energy resolution to give information on chemical state. By measuring the intensities of the XPS peaks for each element and correcting these intensities by relative sensitivity factors (SFs; theoretical tables or standard data), the

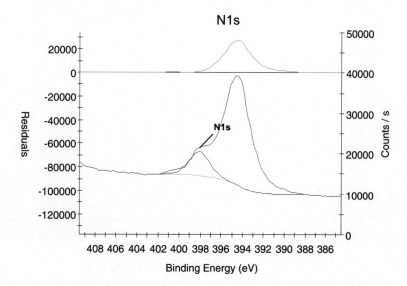

**Figure 8.2**
The N 1s peak (superimposed on the Mo $3p_{3/2}$ peak) sample 3 (ion beam etched at 5 kV for 20 minutes).

composition of the sample can be determined. XPS is a surface-sensitive technique and has an analytical depth of 1−5 nm; consequently, for the analysis of metal-based coatings, it is necessary to remove any surface oxide prior to analyses. This was done using an ion gun attached to the XPS spectrometer. The coating surface is ion-beam etched (at an energy of 5 keV) until the O signal ceased to decrease. XPS analysis of the bulk coating then performed, as in Fig. 8.2.

# 8.6 X-ray diffraction

For completeness in this discussion of techniques, XRD is mentioned briefly, although the process was not used to investigate the samples. All diffraction techniques exploit the scattering of radiation from various sites. X-ray scattering occurs from atoms and molecules in crystals.

Analysis of XRD patterns can help us find the crystal structure of an unknown material. In principle, XRD can provide information on the average spacing between layers or rows of atoms, the orientation of a single crystal or grain, and/or the size, shape, and internal stress of small crystalline regions.

# 8.7 Analysis of results

The composition and coating structure have been examined and analyzed using SEM, TEM, and XPS. SEM was used to examine the surface of the coating and their cross-sectional structures. TEM was used for investigating the microstructure of the coating. XPS was used to give the chemical composition of the various coatings.

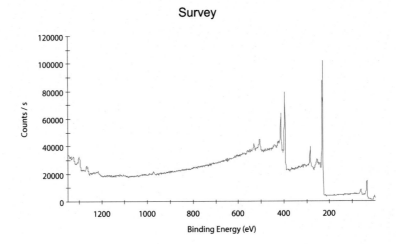

**Figure 8.3**
An XPS survey spectrum taken from sample 4.

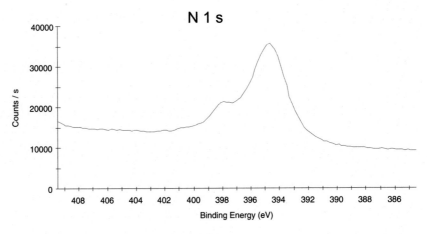

**Figure 8.4**
N1s spectrum—the smaller peak at higher binding energies (overlapping the Mo 3p$_{3/2}$ peak)—sample 4.

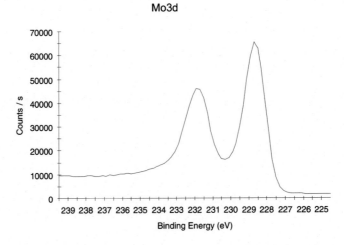

**Figure 8.5**
The Mo3d spectrum—sample 4.

## 8.7.1 XPS data and elemental composition

Strong molybdenum peaks were observed (at 38, 65, 230, 396, 413, and 508 eV) in addition to oxygen (510 eV) and carbon (285 eV). The nitrogen 1 s peak overlapped the Mo $3p_{3/2}$ peak at 396 eV. Oxygen and carbon arose from residual oxide and carbon contamination on the surface after ion etching. Fig. 8.3 and Fig. 8.4 present the N1s and Mo3d spectra, respectively (Fig. 8.5).

**Table 8.1** Compositional data for sample 3.

| Name | Area (P) CPS.eV | SF | At. % | Stoichiometry |
|------|-----------------|------|-------|---------------|
| N1s | 11472.56 | 0.42 | 29.71 | |
| | | | | $Mo_{2.4}N$ |
| Mo3d | 189314.19 | 2.75 | 70.29 | |

**Table 8.2** Compositional data for sample 4.

| Name | Area (P) CPS.eV | SF | At. % | Stoichiometry |
|------|-----------------|------|-------|---------------|
| N1s | 11827.82 | 0.42 | 32.60 | |
| | | | | $Mo_{2.1}N$ |
| Mo3d | 170354.22 | 2.75 | 67.40 | |

**Table 8.3** Compositional data for sample 5

| Name | Area (P) CPS.eV | SF | At.% | Stoichiometry |
|------|-----------------|-------|-------|---------------|
| N1s | 62771.22 | 0.420 | 80.14 | |
| | | | | $Mo_{1/4}N$ |
| Mo3d | 108345.26 | 2.750 | 19.86 | |

**Table 8.4** Compositional data for sample 6

| Name | Area (P) CPS.eV | SF | At. % | Stoichiometry |
|------|-----------------|------|-------|---------------|
| N1s | 9897.58 | 0.42 | 28.30 | |
| | | | | $Mo_{2.5}N$ |
| Mo3d | 174696.37 | 2.75 | 71.70 | |

**Table 8.5** Chemical composition of the various coatings.

|  | At.% Mo | At.% N | Mo/N ratio | Stoichiometry |
|---|---|---|---|---|
| Sample 3 | 70.3 | 29.7 | 2.37 | $Mo_{2.4}N$ |
| Sample 4 | 67.4 | 32.6 | 2.07 | $Mo_{2.1}N$ |
| Sample 5 | 71.7 | 28.3 | 2.53 | $Mo_{2.5}N$ |
| Sample 6 | 20.0 | 80.1 | 0.24 | $Mo_{1/4}N$ |

The XPS peak area is measured as summarized in Table 8.1. The relative SFs are used to calculate the atomic percentages of N and Mo, and from these the stoichiometry can be determined (Table 8.2).

A similar procedure was carried out for samples 3 to 5; the results are summarized in Tables 8.1–8.3, respectively (Table 8.4).

Therefore the chemical composition of the coating, as determined by XPS, is given in Table 8.5.

# 8.8 Crystal structure

In brief, the internal formation structure of matter can be either amorphous or crystallites or both. The strength of a substance, for example, coating, is the result of the type of crystals formation, the orientation of the crystals, and the overall percentage of the formed crystals within the chosen type of coating.

X-ray crystallography (XRD) can be used to determine the arrangement of atoms in crystal as well as the size per unit cell (see Section 8.5). Å is the abbreviation for the unit of length called the Ångström (or Ångstrøm). One angstrom is equal to $10^{-10}$ m (one ten-billionth of a meter or 0.1 nm)

## 8.8.1 Expected structures involving molybdenum and nitrogen

From the stoichiometry results summarized in Table 8.5, considering the Mo–N phase diagram and results of previous work on the MoCuN system (Joseph et al., 2003), we may expect the phase composition of the coatings to be that of Mo(N) and/or $\gamma$-$Mo_2N$.

There is a possibility for other nitrides being present (i.e., other crystallographic forms of $Mo_2N$ or MoN), but these have not been identified in earlier work. Electron diffraction patterns observed in the TEM have been analyzed for four samples (3, 4, 5, and 6). The analysis technique differed slightly as between samples 4 and 5 and samples 3 and 6 (Table 8.6–8.8).

Sample 4 was the most difficult to analyze, but (with reasonable certainty), it proved to be a mixture of Mo and $\gamma$-$Mo_2N$; this analysis is discussed further.

**Table 8.6** X-ray diffraction data for Mo

| Molybdenum | | | |
|---|---|---|---|
| d/Å | $I/I_1$ | hkl | S |
| 2.225 | 102 | 110 | 2 |
| 1.574 | 21 | 200 | 4 |
| 1.285 | 39 | 211 | 6 |
| 1.1127 | 11 | 220 | 8 |
| 0.9952 | 17 | 310 | 10 |
| 0.9085 | 7 | 222 | 12 |
| 0.8411 | 26 | 321 | 14 |
| Sys. | Cubic: | | |
| 3.1472 | Å | | |

**Table 8.7** X-ray diffraction data for $\gamma$-$Mo_2N$.

| $\gamma$-$Mo_2N$ | | | |
|---|---|---|---|
| d/Å | $I/I_1$ | hkl | S |
| 2.404 | 100 | 111 | 3 |
| 2.081 | 48 | 200 | 4 |
| 1.472 | 30 | 220 | 8 |
| 1.255 | 40 | 311 | 11 |
| 1.202 | 12 | 222 | 12 |
| 1.041 | 9 | 400 | 16 |
| 0.955 | 5 | 331 | 19 |
| 0.931 | 69 | 420 | 20 |
| Sys. | Cubic | | |
| 4.163 | Å | | |

Samples 3 and 6 are the easiest to analyzes, and therefore these are discussed first. For reference, we summarize in Tables 8.9 and 8.10, for Mo and $\gamma - Mo_2N$, respectively, the expected lattice-plane spacings (d values), Miller indices (hkl), relative strengths, and values of $S = (h^2 + k^2 + l^2)$. This information is derived from an XRD database.

**Table 8.8** Electron diffraction data for samples 3 and 6.

| R left | R right | R average | d-obs | d-γMN | d-obs/d-γMN | Miller index + |
|--------|---------|-----------|-------|-------|-------------|----------------|
| /pixels | /pixels | /pixels | /Å | /Å | | rel. strength |
| | | | | | | |
| | | | | | | |
| Sample 3: camera constant = | | | | 706.3 | pixel-Å | |
| | | | | | | |
| 291 | 291 | 291.00 | 2.427 | 2.404 | 1.010 | 111−100 |
| 335 | 335 | 335.00 | 2.108 | 2.081 | 1.013 | 200−48 |
| 470 | 472 | 471.00 | 1.500 | 1.472 | 1.019 | 220−30 |
| 555 | 559 | 557.00 | 1.268 | 1.255 | 1.010 | 311−40 |
| | | | | | 1.013 = | average |
| | | | | | | |
| Sample 6: camera constant = | | | | 706.3 | pixel-Å | |
| | | | | | | |
| 282 | 287 | 284.50 | 2.483 | 2.404 | 1.033 | 111−100 |
| 333 | 332 | 332.50 | 2.124 | 2.081 | 1.021 | 200−48 |
| 461 | 470 | 465.50 | 1.517 | 1.472 | 1.031 | 220−30 |
| 553 | 554 | 553.50 | 1.276 | 1.255 | 1.017 | 311−40 |
| | | | | 1.202 | | 222−12 |
| 657 | 658 | 657.50 | 1.074 | 1.041 | 1.032 | 400−9 |
| | | | | | 1.027 = | average |

## 8.8.2 Analysis of samples 3 and 6

From the electron diffraction pattern, ring diameter measurements were made from the electronically recorded diffraction pattern image using computer-based techniques. The ring diameters are expressed in "pixels" and are tabulated in columns 1 to 3 of Table 8.8. Columns 1 and 2 tabulate "left" and "right" values and column 3 the average value.

Using the known value (706.3 pixel Å) for the camera constant, these were then converted to the "observed $d$ values" ($d$-obs) and listed in Column 4. These observed d-values were then compared with the database $d$ values ($d$-db) for $\gamma$-$Mo_2N$, listed in Table 7.6 and reproduced in column 5. Qualitatively, there is a good match for both samples. The ratio $d$-obs/$d$-db was calculated for each ring, and the results are listed in column 6. For each sample the ratio of $d$-obs/$d$-db is fairly consistent,

**Table 8.9** Electron diffraction data for sample 5.

| R, left | R, right | R, average | d-obs | d-γMN | d-obs/ d-γMN | Miller index + |
|---|---|---|---|---|---|---|
| /pixels | /pixels | /pixels | /Å | /Å | | rel. strength |
| Assume camera constant = | | | | 17 | mm | |
| 7.49 | 7.49 | 7.49 | 2.270 | 2.225 | 1.020 | 110−102 |
| 10.49 | 10.48 | 10.49 | 1.621 | 1.574 | 1.030 | 200−21 |
| 12.58 | 12.57 | 12.58 | 1.352 | 1.285 | 1.052 | 211−39 |
| 14.75 | 14.75 | 14.75 | 1.153 | 1.1127 | 1.036 | 220−39 |
| | | | | 0.9952 | | 310−17 |
| 18.00 | 17.96 | 17.98 | | 0.9085 | 1.041 | 222−07 |
| 19.50 | 19.60 | 19.55 | 0.945 | 0.8411 | 1.034 | 321−26 |
| 20.90 | 20.87 | 20.89 | 0.870 | | | |
| 22.30 | 22.20 | 22.25 | 0.814 | | | |
| 23.90 | 23.87 | 23.89 | 0.764 | | | |
| 24.80 | 24.78 | 24.79 | 0.712 | | | |
| 24.70 | 25.60 | 25.15 | 0.686 | | | |
| 26.60 | 26.50 | 26.55 | 0.676 | | | |
| | | | 0.640 | | | |

with average values being about 1.013 for sample 3 and about 1.027 for sample 6.

With sample 6, there is a database line (1.22 Å) that is apparently not present in the sample. But this is expected to be a weak line, and we do not regard its absence as significant. In other respects, there is one-to-one agreement for the strongest expected $\gamma\text{-Mo}_2\text{N}$ lines. It was concluded that $\gamma\text{-Mo}_2\text{N}$ is present in both samples. No other phase was found.

## 8.8.3 Analysis of sample 5

From the electron diffraction pattern, ring diameter measurements were made using ordinary ruler. The ring diameters are expressed in millimeters and are shown in columns 1 to 3 of Table 8.12. Columns 1 and 2 show left and right values and column 3 the average value.

**Table 8.10** Comparison of observed and database lattice-plane spacings for γ-Mo$_2$N.

| d-obs | d-db | d-obs/d-db | d-corrected S | Index + strength | |
|-------|------|------------|---------------|------------------|---|
| 2.599 | 2.404 | 1.081 | 2.396 | 3√ | 111−100 |
| 2.543 | | | 2.345 | | |
| 2.256 | 2.081 | 1.084 | 2.080 | 4√ | 200−48 |
| 1.733 | | | 1.598 | | |
| 1.603 | 1.472 | 1.089 | 1.478 | 8√ | 220−30 |
| 1.360 | 1.255 | 1.084 | 1.254 | 11√ | 311−40 |
| 1.304 | 1.202 | 1.085 | 1.202 | 12√ | 222−12 |
| | Average = | 1.084545 | | | |
| 1.139 | 1.041 | 1.094 | 1.050 | 16√ | 400−09 |
| 1.103 | | | 1.017 | | |
| 0.927 | | | 0.855 | | |
| 0.886 | | | 0.817 | | |
| 0.870 | | | 0.a802 | | |
| 0.758 | | | 0.699 | | |

Using the known value (17 mm) for the camera constant, these were then converted to the d-obs and listed in column 4. These observed d values were then compared with the d-db for γ-Mo$_2$N, listed in Table 8.10 and reproduced in column 5. Qualitatively, there is a good match for the sample. The ratio d-obs/d-db was calculated for each ring, and the results are listed in column 6. For sample 5 the ratio of d-obs/d-db is fairly consistent, with average values being about 1.0355. It was concluded that Mo is present in sample 5 and, at the same time, the calculation showed that there is no γ-Mo$_2$N in this sample.

### 8.8.4 Analysis of sample 4

#### 8.8.4.1 Procedure used for analysis of diffraction data

1. The procedure used to analyze sample 4 is different from those described earlier. During this procedure, it was established that both molybdenum and γ-Mo$_2$N were present, but there was also an anomaly. Ring radii were therefore remeasured using a computer technique, leading to the results for d values summarized in column 1 of Table 8.10.

**Table 8.11** Comparisons with database lattice-plane spacings for Mo.

| d-obs | d-cor | *NOT $\gamma$-Mo$_2$N | d-db(Mo) | d-cor/d-db | Mo | $\gamma$-Mo$_2$N |
|-------|-------|----------------------|----------|------------|-----|------------------|
| 2.599 | 2.396 |                      |          |            |     | 2.404 |
| 2.543 | 2.345 | *& poor              | 2.225    | 1.054      | 110−102$\sqrt{}$ | |
| 2.256 | 2.080 |                      |          |            |     | 2.081 |
| 1.733 | 1.598 | * & weak             | 1.574    | 1.015      | 200−21$\sqrt{}$ | |
| 1.603 | 1.478 |                      |          |            | ?   | 1.472 |
|       |       |                      | 1.285    |            | 211−39 | |
| 1.360 | 1.254 |                      |          |            |     | 1.255 |
| 1.304 | 1.202 |                      |          |            | ?   | 1.202 |
|       |       |                      | 1.113    |            | 220−11 | |
| 1.139 | 1.050 |                      |          |            |     | 1.041 |
| 1.103 | 1.017 | *                    | 0.995    | 1.022      | 310−17$\sqrt{}$ | |
| 0.927 | 0.855 | Weak                 | 0.841    | 1.016      | 321−26$\sqrt{}$ | |
| 0.886 | 0.817 | Weak                 |          |            |     | |
| 0.870 | 0.802 | Weak                 |          |            |     | |
| 0.758 | 0.699 | Weak                 |          |            |     | |

2. Table 8.10 compares experimental and database lattice-plane spacings for $\gamma$-Mo$_2$N. These are tabulated in columns 1 and 2, respectively. Column 3 shows values of the ratio d-obs/d-db. In all cases except line 16 the ratio lies between 1.081 and 1.089, with an average value of 1.0845. This is assumed to be a "correction factor" applicable to all lines because a correction factor of this size is not implausible.

3. This correction factor was then applied to all lines, giving the values in column 4. Comparing the corrected d value with d-db suggests that we can now have reasonable confidence in the identification of the $S = 16$ line. The fit is not as good as for the first five lines, but (as tabulated in column 6) this line strength is relatively weak.

4. The observed and corrected d-values are reproduced in Table 8.11. An asterisk in column 3 indicates those lines that definitely cannot be easily fitted as a $\gamma$-Mo$_2$N line.

5. An attempt was then made (in Table 8.14) to fit the Mo lines to the corrected d values. The strongest Mo lines are tabulated in column 4, matched to the corrected d values. Although the fit is not particularly good, all the lines so far unidentified can be attributed to Mo, and almost all the

**Table 8.12** Comparisons with database lattice-plane spacings for other nitrides.

| d-obs | d-cor | *NOT γ-Mo$_2$N | Nitride 2 | | Nitride 5 | | γ-Mo$_2$N |
|---|---|---|---|---|---|---|---|
| 2.599 | 2.396 | | 2.396 | | 2.370 | | 2.404 |
| 2.543 | 2.345 | *Poor | | | 2.370 | | |
| 2.256 | 2.080 | | 2.106 | | 2.090 | ? | 2.081 |
| | | | 2.012 | ? | 2.002 | | |
| 1.733 | 1.598 | *Weak | | | | | |
| 1.603 | 1.478 | | 1.487 | | 1.482 | ? | 1.472 |
| | | | 1.453 | ? | 1.447 | | |
| 1.360 | 1.254 | | 1.261 | | 1.261 | | 1.255 |
| 1.304 | 1.202 | | 1.195 | | 1.192 | | 1.202 |
| 1.139 | 1.050 | | | | | | 1.041 |
| 1.103 | 1.017 | * | | | | | |
| | | | 0.940 | ? | | | |
| 0.927 | 0.855 | Weak | | | | | |
| 0.886 | 0.817 | Weak | | | | | |
| 0.870 | 0.802 | Weak | | | | | |
| 0.758 | 0.699 | Weak | | | | | |

expected (from the database) strong Mo lines are present (although in some cases, they overlap with γ-Mo$_2$N lines). There is some difficulty in identifying the expected 1.285 Å and 1.113 Å Mo lines with the observed lines, but the presence of identified γ-Mo$_2$N lines close to these values may be making this difficult.

6. Column 4 shows the ratio of the corrected d value to the database d value, for the postulated Mo lines. All the postulated Mo lines have d-db less than the corrected d values, which might indicate the existence of strained crystallites, but the evidence is not strong enough to make this a firm conclusion.

7. Columns 5 and 6 show the d values corresponding to two other forms of nitride listed in the XPS database. These materials are less able to explain the unidentified lines than Mo is.

   Although for both nitrides there are expected lines that are not present in the observed data, the closeness of the other listed values to the γ-Mo$_2$N values means that it is difficult is to totally exclude the possibility that one or both of these nitrides are present, in addition to the

$\gamma$-Mo$_2$N. We emphasize, however, that $\gamma$-Mo$_2$N is a significantly better match to the diffraction data than are the other forms of nitride (Table 8.12)

8. Thus, conclusions from the electron diffraction analysis of specimen 4 are as follows:

1. Nanocrystallites of $\gamma$-Mo$_2$N are most likely present as the majority phase.
2. There is a second phase present that is most probably molybdenum, in which case the Mo phase may conceivably be in a slightly strained state.
3. It seems difficult to totally exclude the possibility that nanocrystallites of other forms of nitride may also be present, but there is no convincing evidence for their presence.

# References

Cárdenas-Lizana, F., Gómez-Quero, S., Perret, N., Kiwi-Minsker, L., & Keane, M. (2010). b-Molybdenum nitride: synthesis mechanism and catalytic response in the gas phase hydrogenation of p-chloronitrobenzene. *Catalysis Science and Technology, 2011*(1), 794–801. <https://pubs.rsc.org/en/content/articlepdf/2011/cy/c0cy00011f>.

Chaudhari D., Patil T., Raichurkar P., & Daberao A. (2018). *Review on nanotechnology & its application in coating industry.* <https://www.researchgate.net/publication/325380668_Review_on_Nanotechnology_Its_Application_in_Coating_Industry>.

Humphreys C., (2002). Theory of electron scattering and electron diffraction. Chapter 2.9.1. In *Scattering and inverse scattering in pure and applied science* (pp. 1287–1303). ScienceDirect. <https://reader.elsevier.com/reader/sd/pii/B9780126137606500693?token=488B087C08A8914826E4F5CAC93AA50DE4194AFAF1827AC6AAD638325805BF0F8DCA7E0B7E8A51F22C1DE1C184FAD3D8>.

Joseph M., Tsotsos C., Leyland A., Mathews A., Baker M., Kench P., & Rebholz C. (2003). *Characterisation and tribological evaluation of nitrogen-doped molybdenum-copper PVD metallic nanocomposite films.* University of Hull, University of Surrey, Robert Bosch (Germany).

# Tests and training

The principal objective of this chapter is the development and dissemination of a set of procedures that enable the characteristics of any machine olfaction device (MOD) to be quantifiably assessed.

This depends on the quality of the software program compiled or the artificial intelligence (AI) method of training introduced to the system.

If the device is specifically designed for only one type of application, for example, detection of explosives, then obviously one criteria of testing will be implemented; however, if the device is required for a variety of applications, then there will be a number of criteria needed to be implemented.

To check and demonstrate the MOD device functionalities and capability in detecting gases and odors, a selection of various testing samples has been allocated as part of the initial training for the device, specifically for the metal-oxide semiconductor (MOS) sensor, as an example. Further testing for the biosensor, not included in this chapter, should be provided as another test example for the MOD device.

For the MOS sensor, the work presented confirms that the MOD responded accordingly in experiment. In many cases, when compared with previous work carried out in the same field, the result is as expected, that is, accurate output.

The detection of the desired compound in each sample allows judgment to be passed upon the state of the sample, in relation to both the constant and variable conditions during each experiment. That means the response of the sensor is clearly dependent on the volatility and concentration of the target component, as well as the functional condition of the MOD.

The need for further work on sampling techniques is needed, as improved MOD software and better hardware functionality will allow a significant increase in sample headspace identifications, as well as achieving more rapid results and higher accuracy.

However, for comprehensives training, using Java, Python, Lisp, Prolog, or C++ are some of the methods of training MOD without the use of programmed coding.

Among these languages, Python is the most popular among AI developer.

Introduction to Machine Olfaction Devices.
DOI: https://doi.org/10.1016/B978-0-12-822420-5.00013-1

# 9.1 The device

The general basic outline for the MOD, previously mentioned in the Chapter 1: Backgrounds, materials and process, is to be used to detect and recognize odors/vapors, either with two or more of the electronic chemical sensors and biosensors, with partial specificity and appropriate pattern recognition system, or with an AI program, that is, without the command average computer software program (see Section 9.6). The use of AI is to enable the device to recognize simple or complex odors/or gases by applying specifically deep learning (DL) methods.

## 9.1.1 Basic characterizations

The objective is to develop standard procedures and protocols to enable quantitative characterization of MOD instrumentation. Specific parameters to be addressed include repeatability, reproducibility, uncertainty, range, sensitivity, and traceability of the measurements, together with associated calibration.

Other important areas, which should also be considered, are as follows:

1. Detection limit—how small is one substance sample without a change in the response.
2. The detection of an analyte in a variable sample (e.g., detecting the same element in different samples that contains different substances).
3. Standardization (or lack of it—the need for the development of standardized procedures and protocols that enable objective's quantitative and qualitative comparisons to be made between different types of devices).

There are other factors as well, including engineering and technical problems, which need to be looked at in detail, such as power consumption (for mobile MOD) and new modes of operations. These two engineering design problems can be solved using micro-hotplate (see Section 6.5.3, Chapter 6: Microsystem).

## 9.1.2 Important experimental tests

*Repeatability*—closeness of the agreement between the results of successive measurements of the same variable, carried out under the same conditions.

*Reproducibility*—closeness of the agreement between the results of measurements of the same variable, carried out under changed conditions of measurement.

*Uncertainty of measurement*—a parameter associated with the result of a measurement.

*Sensitivity*—change in the response of a measuring instrument divided by the corresponding change in the stimulus.

*Measuring range*—set of values of measure and for which the error of a measuring instrument is intended to lie within specified limits.

*Traceability*—property of the result of a measurement or the value of a standard whereby it can be related to stated references.

### 9.1.3 Experimental work

Various characteristics of MOD experimental performance are as follows:

*Drift*—Sensor drift needs to be monitored and/or compensated for, in order to achieve good reproducibility.

*Comparability of sensors*—Particularly in terms of comparability before and after sensor replacement.

*Environmental influence*—The effects of temperature, pressure, and humidity need to be quantified, in order to define operational limits for a given level of performance.

*Realization of suitable calibration artifacts*—At present, the majority of MOD measurements are based on application-specific comparison measurements; therefore the specific chemicals that are in any mixture are not usually known.

## 9.2 Samples

The following samples have been chosen to start basic experiments on the MOD device. Various concentrations of aqueous acetone, ethanol, and methanol have been used, as well as HPLC water (HPLC = High Performance Liquid Chromatography).

1.  Aqueous acetone, ethanol, and methanol concentrations
    0.00% (water) 0.05% 0.10% 1% and 10%
2.  Time in water bath
    From 1−27 minutes
3.  Temperature variations
    From 33.1°C−60°C
4.  Changing the sensor's resistance

Variations of sensor's resistance can be made during each experiment using the control knob on the MOD device. One unit on the graph is equal to 10 'Ω (Ohm).

## 9.3 Applying methodology to the samples

There are four steps in this methodology, these are:

Sample concentration.
Time (related to water bath, injection, change in sensor's resistance).
Temperature (related to water bath).
Sensor resistance.

Using tables of MOS resistance rating, the experiments are discussed further in the following pages.

Each experiment is repeated three times under the same conditions. Each set of tests have one or more variations applied to it during each experiment.

It is important to analyze the relationships between the variables for each sample provided in the laboratory as it will indicate the direction of changes (if any) when these variables are applied to the system generally, and particularly on the MOS.

### 9.3.1 Instrumentation

The analysis was conducted using MOD, PC (with Matlab software), water bath (with heater), timer, measuring devices, and an injector.

The MOD was kept switched on throughout the duration of the experiments to ensure that the sensor stabilization was unchanged and ready for the next session of experiments (as it takes 3 days to stabilize the sensor if the device is switched off. (Figs. 9.1 and 9.2)

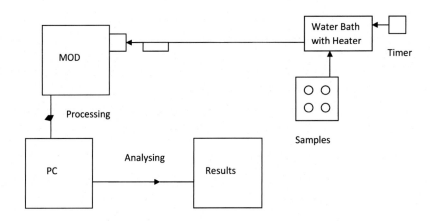

**Figure 9.1**
Simple illustrations for various steps on each experiment.

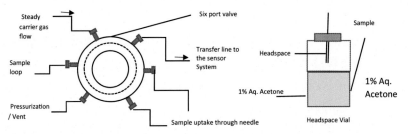

**Figure 9.2** Schematic diagrams showing the signal derived from sample concentration and headspace sampler as used in all the experiments in this work. *Source: Redrawn from Strathamann, 2001.*

## 9.3.2 Experiment layout

All the sensor's resistance measurements for these experiments are above 125 KΩ (baseline resistance for the sensor). As such, what is measured here is the additional resistance produced when the sensor is exposed to the gaseous headspace.

The following are two examples of tests conducted in the laboratory. These tests consist of a number of experiments with a brief comment on each result (as noted from the actual laboratory logbook).

## 9.3.3 Test 1

The only variable in the following experiments is the concentrations of each element in each sample. The rest of the factors are kept constant.

### 9.3.3.1 A1

| Aqueous methanol | Sample concentration: 0.05% v/v | MOS resistance rating 60 Ω |
|---|---|---|
| | Temperature: 25°C | |
| | Time in water bath: 5 minutes | |
| | Injection speed: unchanged | |
| | Sensor resistance: unchanged | |

The MOD sensor resistance in Graph 9.A1 shows that the sensor response took approximately less than 1 second to respond to the headspace gas. The resistance curve declined slowly and ends above the time scale row (X-axis), possibly

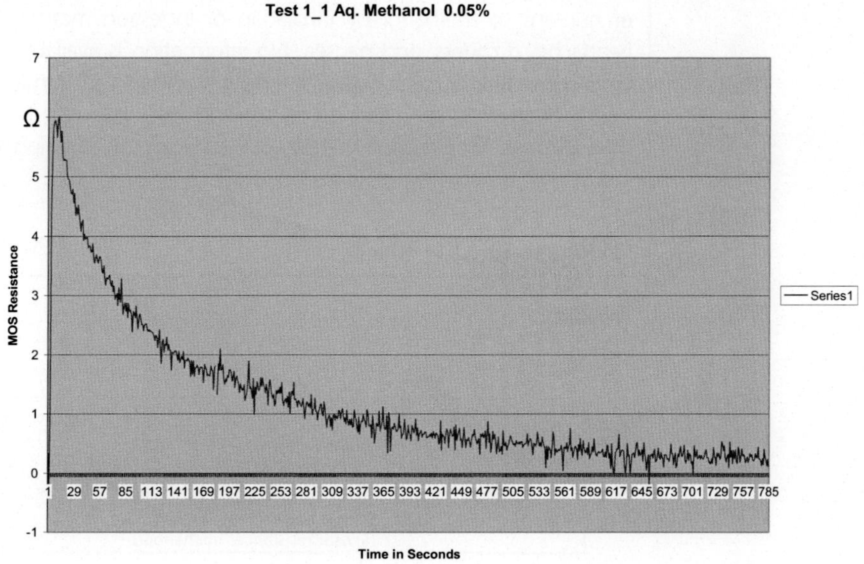

**Test 1_1 Aq. Methanol 0.05%**

**Graph 9.A1**

indicating that there is trace of similar or different samples in the chamber prior to this experiment. Therefore the function of the device is normal.

According to the scale of the Graph 9.A1, the sensor response reaches up to 60 $\Omega$. (Box 9.1)

---

## Box 9.1 Methanol

Methanol appears as a colorless fairly volatile liquid with a faintly sweet pungent odor like that of ethyl alcohol. Completely mixes with water. The vapors are slightly heavier than air and may travel some distance to a source of ignition and flash back. Any accumulation of vapors in confined spaces, such as buildings or sewers, may explode if ignited. Used to make chemicals, to remove water from automotive and aviation fuels, as a solvent for paints and plastics, and as an ingredient in a wide variety of products.

Methanol is the primary alcohol that is the simplest aliphatic alcohol, comprising a methyl and an alcohol group. It has a role as an amphiprotic solvent, a fuel, a human metabolite, an *Escherichia coli* metabolite, a mouse metabolite, and a *Mycoplasma genitalium* metabolite. It is an alkyl alcohol, a one-carbon compound, a volatile organic compound, and a primary alcohol. It is a conjugate acid of a methoxide.

Methanol is released to the environment during industrial uses and naturally from volcanic gases, vegetation, and microbes. Exposure may occur from ambient air and during the use of solvents. Acute (short-term) or chronic (long-term) exposure of humans to methanol by inhalation or ingestion may result in blurred vision, headache, dizziness, and nausea. No information is available on the reproductive, developmental, or carcinogenic effects of methanol in humans. Birth defects have been observed in the offspring of rats and mice exposed to methanol by inhalation. EPA has not classified methanol with respect to carcinogenicity.

*PubChem (undated)*

National Centre for Biotechnology information

---

*9.3.3.2 B1*

| Aqueous methanol | Test 1.2<br>Sample concentration: 1% v/v<br>Temperature: 25°C<br>Time in water bath: 5 minutes<br>Injection speed: unchanged<br>Sensor resistance: unchanged | MOS resistance rating 68 $\Omega$ |
| --- | --- | --- |

When the concentration changed to 1%, the increase can be noticed by 68 Ω on the graph. The same experiment was repeated three times but showed no significant changes. (Graph 9.B1)

Using tables with MOS resistance rating, discussions for the experiments can be addressed in the following manner:

| Aqueous methanol | Sample concentration: 0.05%<br>Time in water bath at room temperature:<br>5 minutes<br>Injection speed: unchanged<br>Sensor resistance: unchanged | MOS resistance rating 6 units |
| --- | --- | --- |

The MOD has shown that the resistances reached to the scale of 6 under the conditions described in the above table. The experiment has been repeated

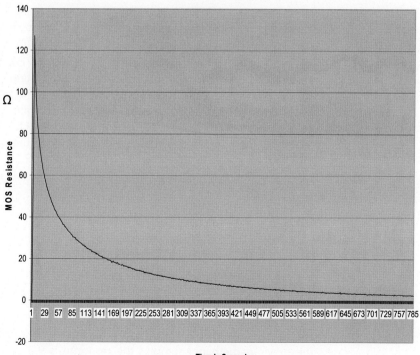

**Test1_2 Aq. Methannol 1%**

**Graph 9.B1**

three times under the same conditions and on each occasion the result obtained was the same.

| Aqueous methanol | Test 1.2<br>Sample concentration: 1%<br>Time in water bath at room temperature: 5 minutes<br>Injection speed: unchanged<br>Sensor resistance: unchanged | MOS resistance rating 6.9 units |
|---|---|---|

When the concentration changed to 1% then the increase can be noticed by 10 units on the graph that is, 10 units above the point 6. The same experiment was repeated three times and no significant changes noticed.

*9.3.3.3 C1*

| Aqueous methanol | Test 1.3<br>Sample concentration: 0.1%<br>Time in water bath at room temperature: 5 minutes<br>Injection speed: unchanged<br>Sensor resistance: unchanged | MOS resistance Rating 7.5 units |
|---|---|---|

The resistances reached to the scale of 7.5 under the conditions described in the above table. The experiment has been repeated three times under the same conditions and on each occasion the result obtained was the same. The response here was much higher than the previous higher concentration of aqueous methanol. The only explanation can be found here is that the MOD device was moved from its original place by another member of the staff. All the other devices within the MOD are still the same and in working order. (Graph 9.C1)

*9.3.3.4 D1*

| Aqueous ethanol | Test 1.4<br>Sample concentration: 1%<br>Time in water bath at room temperature: 5 minutes<br>Injection speed: unchanged<br>Sensor resistance: unchanged | MOS resistance rating 9.3 units |
|---|---|---|

For concentration of 1.4% of aqueous ethanol under the same conditions as the previous methanol samples the response is 9.3. The MOD in these three repeated experiments for ethanol 1.4% is responding (i.e., increasing current resistance) much quicker as the Graph 9.D1 illustrates (Box 9.2).

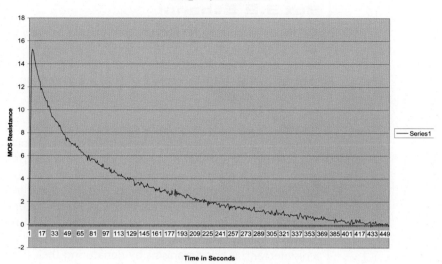

**Graph 9.C1**

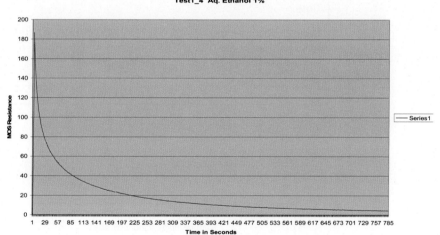

**Graph 9.D1**

*9.3.3.5 E1*

| Aqueous acetone | Test 1.4 | MOS resistance rating |
|---|---|---|
| | Sample concentration: 1% | |
| | Time in water bath at room temperature: 5 minutes | |
| | Injection speed: unchanged | |
| | Sensor resistance: unchanged | |

## Box 9.2 Ethanol

The chemical compound *ethanol*, also known as *ethyl alcohol* or *grain alcohol*, is the bio-alcohol found in alcoholic beverages. When nonchemists refer to "alcohol," they almost always mean ethanol. It is also increasingly being used as a fuel (usually replacing or complementing gasoline). Ethanol's chemical formula is $C_2H_5OH$.

*Properties*

Pure ethanol is a flammable, colorless liquid with a boiling point of 78.5°C. Its low melting point of −114.5°C allows it to be used in antifreeze products. It has a pleasant odor reminiscent of whiskey. Its density is 789 g/l about 20% less than that of water. It is easily soluble in water and is itself a good solvent, used in perfumes, paints, and tinctures. Alcoholic drinks have a large variety of tastes, since various flavor compounds are dissolved during brewing. A solution of 70%–85% of ethanol is commonly used as a disinfectant; it kills organisms by denaturing their proteins and dissolving their lipids: it is effective against most bacteria and fungi, and many viruses, but is ineffective against bacterial spores. This disinfectant property of ethanol is the reason that alcoholic beverages can be stored for a long time.

Ethanol can lose a proton from the hydroxyl group and is a very weak acid, weaker than water. The CAS number of ethanol is 64−17−5 and its UN number is UN 1170.

*World of Molecules (undated)*

(Graph 9.E1)

*9.3.3.6 F1*

| HPLC water | Test 1.6 | MOS resistance rating below zero |
|---|---|---|
| | Sample concentration: 0% v/v | |
| | Temperature: 25°C | |
| | Time in water bath: 5 minutes | |
| | Injection speed: unchanged | |
| | Sensor resistance: unchanged | |

In the course of the three experiments with HPLC water, the response noted usually starts and ends below zero. Four experiments were performed on this sample but only one resulted in the graph (Graph 9.F1) on zero scale. The

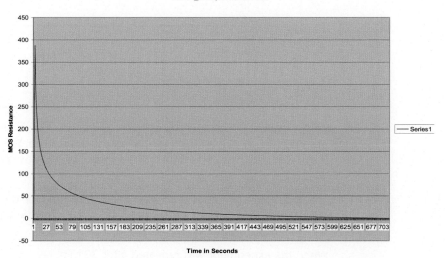

**Graph 9.E1**

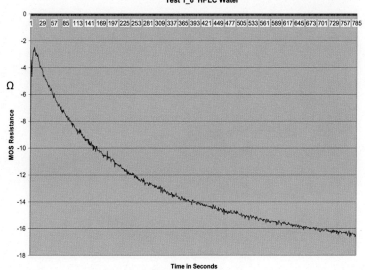

**Graph 9.F1**

explanation given is that, getting results below zero may indicate that the injection of the sample was performed too fast. The experiment was then repeated with a slower pace of injection. However, the result remained the same.

### 9.3.3.7 G1

| | | |
|---|---|---|
| Aqueous ethanol | Test 1.7 | MOS resistance rating 46 Ω |
| | Sample concentration: 0.1% v/v | |
| | Temperature: 25°C | |
| | Time in water bath: 5 minutes | |
| | Injection speed: unchanged | |
| | Sensor resistance: unchanged | |

With this test, the aqueous ethanol concentration was 0.1%. The graph obtained from this experiment crossed below the Y-axis at 309 seconds (i.e., went below zero) after the sensor was exposed to the headspace gas.

This indicates that either there was insufficient time for the device to clear the chamber from the traces of the previous sample, or the injection was too fast (as mentioned previously) or both.

When the experiment was repeated after 30 minutes (to ensure better clearing of the chamber), some minor changes were noted.

In the illustration (Graph 9.G1), the scale for the sensor response reached 46 Ω on the graph.

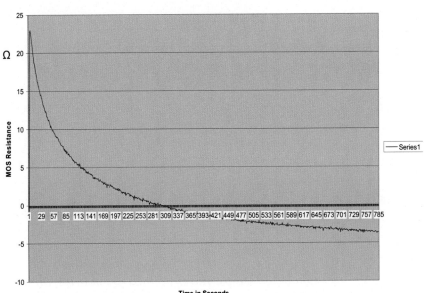

Test 1_7 Aq. Acetone 0.1%

**Graph 9.G1**

*9.3.3.8 H1*

| | | |
|---|---|---|
| Aqueous acetone | Test 1.8 | MOS resistance rating 66 $\Omega$ |
| | Sample concentration: 0.05% v/v | |
| | Temperature: 25°C | |
| | Time in water bath: 5 minutes | |
| | Injection speed: unchanged | |
| | Sensor resistance: unchanged | |

Graph 9.H1 shows that even after setting the device aside for 30 minutes, to clear the chamber, there were still traces of the previous samples in the device.

The idea was to set aside the device for another 1–2 hours before the next experiment to allow sufficient time for the thorough clearing of the chamber (*The device itself automatically cleans the chamber by pumping zero grade air into the chamber for few minutes after each experiment*).

After 448 seconds, the graph crossed below 0 on the Y-scale (MOS resistance).

The scale in the graph is 6.6 units of '$\Omega$ (66 '$\Omega$).

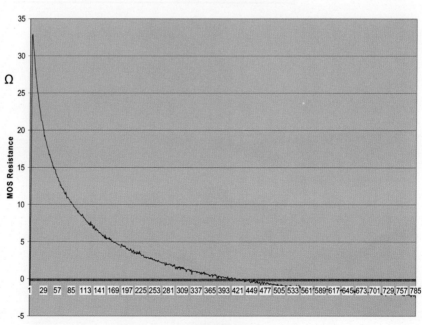

**Test 1_8  Aq. Acetone  0.05%**

**Graph 9.H1**

.

*9.3.3.9 I1*

| | | |
|---|---|---|
| Aqueous acetone | Test 1.9 | MOS resistance rating 80 Ω |
| | Sample concentration: 0.1% v/v | |
| | Temperature: 25°C | |
| | Time in water bath: 5 minutes | |
| | Injection speed: unchanged | |
| | Sensor resistance: unchanged | |

After the two hours of chamber clearing, the expected result was obtained—that is, the graph no longer starts from below zero.

All the three experiments showed the same result of 80 Ω of sensor response. (Graph 9.I1)

*9.3.3.10 J1*

| | | |
|---|---|---|
| Aqueous ethanol | Test 1.10 | MOS resistance rating 79 Ω |
| | + Sample concentration: 0.05% v/v | |
| | Temperature: 25°C | |
| | Time in water bath: 5 minutes | |
| | Injection speed: unchanged | |
| | Sensor resistance: unchanged | |

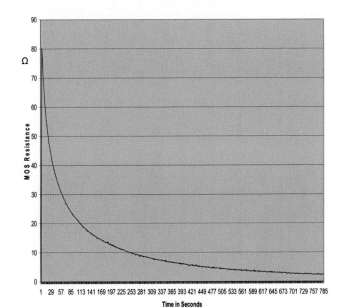

**Graph 9.I1**

Test 1.10 showed a slower response (in relation to the previous experiments). As compared to the abrupt upward gradient of the graph after the change in the resistance of the sensor, its descent was more gradual and rugged.

The sensor response here is 79 $\Omega$ on the graph scale. (Graph 9.J1)

### 9.3.3.11 K1

Graph 9.K1 shows all the experiments for the different samples in Test 1. The 10% concentration of all the different elements reached the highest peak

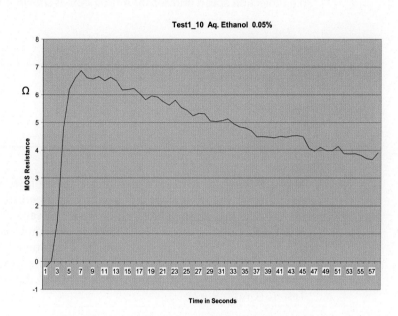

**Graph 9.J1**

.

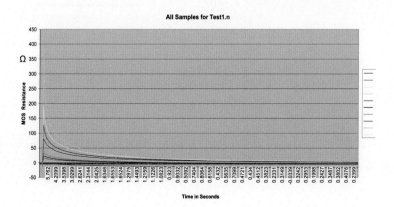

**Graph 9.K1**

.

(obviously due to the very high concentration) on the graph (88 Ω). The lowest is the HPLC water.

### 9.3.4 Test 2

There are two variables in this test. The variables are:

1. The duration of time that each sample was placed in the water bath.
2. The different concentrations in water for the three elements.

The temperature was kept constant at 40°C together with the other variables (i.e., injection speed and sensor resistance), as shown in the following tables.

### 9.3.4.1 A2

| Aqueous methanol | Test 7.1<br>Sample concentration: 0.05% v/v<br>Temperature: 40°C<br>Time in water bath: 15 minutes<br>Injection speed: unchanged<br>Sensor resistance: unchanged | MOS resistance rating 38 Ω |
|---|---|---|

Test 2.1 shows a resistance reaching a scale of 38 Ω under the conditions described in the table above. It seems that when there is a rise in the temperature of the sample, the response of the sensor increases accordingly. The time scale for the sensor therefore was much shorter when the temperature was 25°C. This can also be explained by the longer duration of time that the sample was placed in the water bath—that is, for 15 minutes, instead of the usual 5 minutes during the experiments of Test 1. (Graph 9.A2)

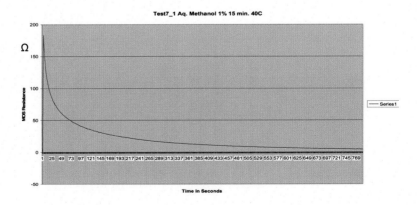

**Graph 9.A2**

.

### 9.3.4.2 B2

| | | |
|---|---|---|
| Aqueous methanol | Test 7.2 | MOS resistance rating 86 Ω |
| | Sample concentration: 10% v/v | |
| | Temperature: 40°C | |
| | Time in water bath: 1 minute | |
| | Injection speed: unchanged | |
| | Sensor resistance: unchanged | |

By heating the sample for less than 1 minute in the water bath, the odor concentration (headspace) is reduced. This is because the MOD has managed to remove what has been left from the sample (in the chamber) faster (as the graph illustrates) than the previous experiment. However, the response of the sensor remains the same, regardless of the duration of the time spent in the water bath.

The sensor's resistances reached to 86 Ω. (Graph 9.B2)

### 9.3.4.3 C2

| | | |
|---|---|---|
| Aqueous acetone | Test 7.3 | MOS resistance rating 44 Ω |
| | Sample concentration:0.05% | |
| | Temperature 40°C | |
| | Time in water bath: 10 minutes | |
| | Injection speed: unchanged | |
| | Sensor resistance: unchanged | |

Despite setting aside the device for 30 minutes in order to clear the chamber, the curve still dipped below zero in the time scale. The response of the sensor on the same time scale was almost similar to the previous experiment. After 530 seconds, the graph crossed below 0 on the Y-scale (MOS resistance). The sensor's resistances reached 44 Ω. (Graph 9.C2) (Box 9.3)

### 9.3.4.4 D2

| | | |
|---|---|---|
| Aqueous ethanol | Test 7.4 | MOS resistance rating 78 Ω |
| | Sample concentration: 0.1% v/v | |
| | Temperature: 40°C | |
| | Time in water bath: 3 minutes | |
| | Injection speed: unchanged | |
| | Sensor resistance: unchanged | |

The problem continued during this experiment with the curve dipping below zero on the time scale, even though the device was left unused for 2 hours. It

Test 7_2  Aq. Methanol 10%  1 min.  40C

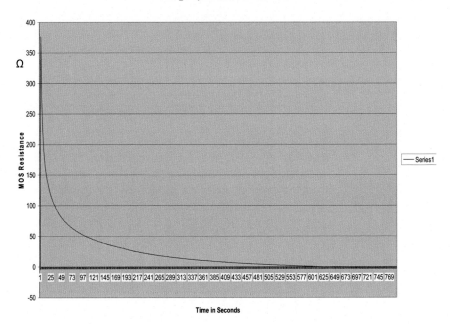

**Graph 9.B2**

Test 7_3  Aq. Acetone  0.05%  10 min.  40C

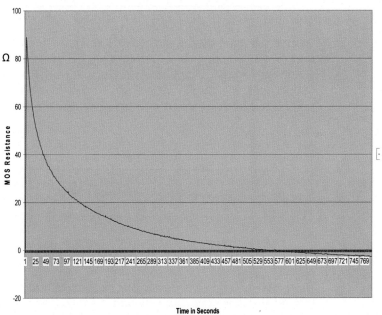

**Graph 9.C2**

## Box 9.3 Acetone (propanone)

Acetone, a colorless liquid also known as propanone, is a solvent used in manufacture of plastics and other industrial products. Acetone may also be used to a limited extent in household products, including cosmetics and personal care products, where its most frequent application would be in the formulation of nail polish removers. Acetone occurs naturally in the human body as a by-product of metabolism. Acetone is a primary ingredient in many nail polish removers. It breaks down nail polish, making it easy to remove with a cotton swab or cloth. It is widely used because it can easily mix with water and evaporates quickly in the air. Acetone is widely used in the textile industry for degreasing wool and degumming silk. As a solvent, acetone is frequently incorporated in solvent systems or "blends," used in the formulation of lacquers for automotive and furniture finishes. Acetone also may be used to reduce the viscosity of lacquer solutions.

Acetone has been extensively studied and is generally recognized to have low acute and chronic toxicity. The United States Food and Drug Administration has determined acetone is safe for use as an indirect food additive in adhesives and food-contact coatings and is regarded as a "Generally Recognized as Safe" substance at certain concentrations.

Acetone has undergone a comprehensive review under the United States Environmental Protection Agency's (EPA) Voluntary Children's Chemical Evaluation Program (VCCEP). The objective of VCCEP was to ensure that adequate toxicity and exposure information was available to assess potential risks to children. This VCCEP review of acetone included a hazard assessment, an exposure assessment, and a risk characterization.

Acetone has undergone regulatory and scientific evaluations under the European Chemical Agency REACH program, the World Health Organization International Programme on Chemical Safety, and EPA's Integrated Risk Information System assessment program.

In a draft screening assessment for health and environmental effects, Environment Canada identified no critical health effects from exposures expected to occur from occasional, intermittent use of certain products containing acetone.

The United States Occupational Safety & Health Administration sets safe workplace exposure limits. Workplaces where acetone is used, such as nail salons, can keep exposure levels below safety levels by using proper ventilation and following manufacturer's instructions.

Acetone is highly flammable but is generally recognized to have low acute and chronic toxicity. If inhaled, acetone could cause a sore throat or cough.

*ChemicalSafteyFacts.org (undated)*

was suggested using 10% concentrations—which may be the likely cause of the problems. The device is highly sensitive and as such, anything above 1% would cause a problem for the other experiments, as traces of the highly concentrated samples will stay in the device for a longer period.

After 337 seconds, the graph crossed below 0 on the Y-scale (MOS resistance).

The sensor's resistances reached to the scale of 78 Ω. (Graph 9.D2)

### 9.3.4.5 E2

| | | |
|---|---|---|
| HPLC water | Test 7.5 | MOS resistance rating 22 Ω |
| | Sample concentration: 0.0% v/v | |
| | Temperature: 40°C | |
| | Time in water bath: 16 minutes | |
| | Injection speed: unchanged | |
| | Sensor resistance: unchanged | |

The response is from unit $+1$ ($+0.5$ on the graph scale) down to unit $-6$ ($-2.98$ on the graph scale) and even though the sensor response remained unchanged within the first few units, above the Y-axis; the remaining two-thirds of the curve went below zero.

The difficulty for the sensor to respond to zero traces (i.e., to the water sample) is clearly illustrated in the graph (Graph 9.E2), as well as in the previous experiments performed on the same sample. These, however, were under different conditions.

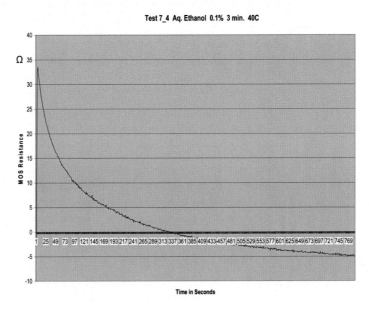

Test 7_4 Aq. Ethanol 0.1% 3 min. 40C

**Graph 9.D2**

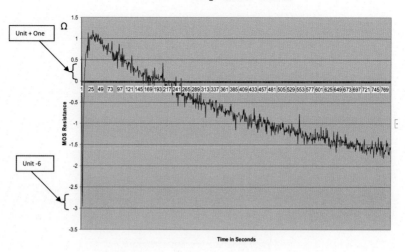

**Graph 9.E2**

After 216 seconds, the graph crossed below 0 on the Y-scale (MOS resistance).

The sensor's resistances reached to the scale of 22 Ω.

*9.3.4.6 F2*

| Aqueous acetone | Test 7.6 | MOS resistance rating 48 Ω |
|---|---|---|
| | Sample concentration: 1% v/v | |
| | Temperature: 40°C | |
| | Time in water bath: 20 minutes | |
| | Injection speed: unchanged | |
| | Sensor resistance: unchanged | |

Test 7.6 shows a resistance reaching a scale of 48 Ω, as expected compared with the other two similar experiments. This is a normal average sensor response for aqueous acetone at 1% concentration.

The sensor's resistance reached a scale of 48 Ω. (Graph 9.F2)

*9.3.4.7 G2*

| Aqueous acetone | Test 7.7 | MOS resistance rating 79 Ω |
|---|---|---|
| | Sample concentration: 0.1% v/v | |
| | Temperature: 40°C | |
| | Time in water bath: 1 minute | |
| | Injection speed: unchanged | |
| | Sensor resistance: unchanged | |

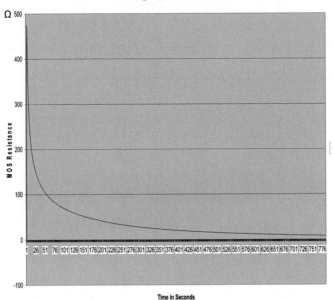

**Graph 9.F2**

Test 7.7 shows a resistance reaching a scale of 79 Ω.

After 384 seconds, the graph crossed below 0 on the Y-scale (MOS resistance). The reasons for this reaction are similar to that of the previous experiments. (Graph 9.G2)

## 9.3.4.8 H2

| Aqueous methanol | Test 7.8<br>Sample concentration: 0.1% v/v<br>Temperature: 40°C<br>Time in water bath: 1 minute<br>Injection speed: unchanged<br>Sensor resistance: unchanged | MOS resistance rating 79 Ω |
|---|---|---|

Graph 9.H2 gives a good indication that the result of this experiment is normal and as expected. Both the response and the decline of the curve on the time scale are similar to that in all the previous three experiments performed on the same sample, under the same conditions.

The sensor's resistance reached a scale of 79 Ω.

Test 7_7  Aq. Acetone 0.1%  1 min.  40C

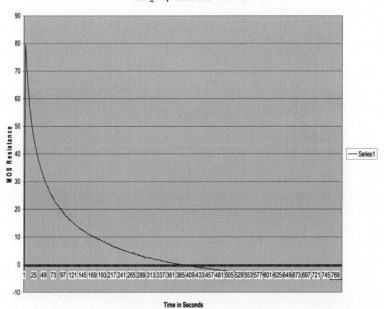

**Graph 9.G2**

Test 7_8  Aq. Methanol  0.1%  12 min.  40C

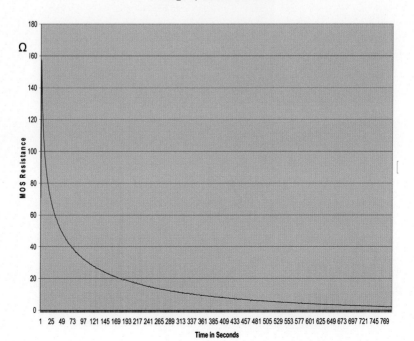

**Graph 9.H2**

*9.3.4.9 12*

| | | |
|---|---|---|
| Aqueous ethanol | Test 7.9 | MOS resistance rating 39 Ω |
| | Sample concentration: 1% v/v | |
| | Temperature: 40°C | |
| | Time in water bath: 30 seconds. | |
| | Injection speed: unchanged | |
| | Sensor resistance: unchanged | |

In this experiment, the sensor response is slightly slower than those carried out under similar conditions, for different elements. The reason is that the sample remained for only 30 seconds in the water bath, which may not be long enough for the sample to be heated sufficiently for the experiment.

The response is 39 Ω on the graph scale (Graph 9.12).

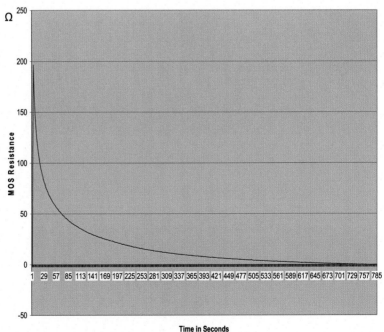

**Test 7_9  Aq. Ethanol  1% 30 Sec.  40C**

**Graph 9.12**

*9.3.4.10 J2*

| | | |
|---|---|---|
| Aqueous methanol | Test 7.10 | MOS resistance rating 32 Ω |
| | Sample concentration: 0.05% v/v | |
| | Temperature: 40°C | |
| | Time in water bath: 3 minutes | |
| | Injection speed: unchanged | |
| | Sensor resistance: unchanged | |

Graph 9.J2 is a good illustration of the device's technical deficiency in relation to the clearing of the chamber. In addition, this may indicate that the age of the sensor could be the contributing factor to this response (as the sensor should be replaced every 3 months).

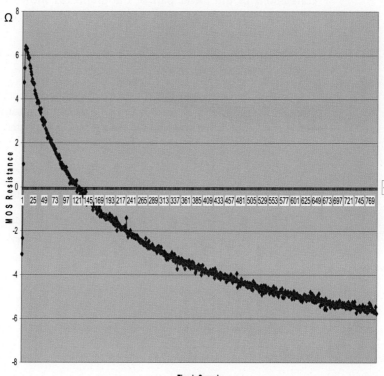

**Graph 9.J2**

## 9.4 Data analysis

The experiments above indicate a clear relationship between the sample concentration and the sensor resistance. The sensor response (current resistance) increased as the concentration in the sample increased (and the opposite can be true, i.e., less concentration = less current resistance).

This means that the resistance change is caused by a loss or gain of surface electrons because of absorbed oxygen reacting with the target vapor (headspace). If the oxide is an n-type, there is either a donation or subtraction of electrons from the conducting band (Fig. 9.3).

Therefore the MOS sensor can be made quantitative, as the magnitude of change to electrical resistance is a direct measure of the concentration of the target analyte present.

From the information gathered from the various tests (each containing 10 experiments and each experiment repeated three times), the relatively close correlations of the lower part of the graph (as shown in Graph 9.A3) mostly reflect on the response of the sensor when lower concentrations of the three elements were used.

All the samples were concentrated from approx. 0%–10% (0.00% "water" 0.05% 0.10% 1% and 10%).

The correlation between MOS resistance and various concentrations of methanol, ethanol, acetone, and water has been plotted in the same graph (Graph 9.A3) and it can be said that the result was "nearly" as expected, that is, concentration

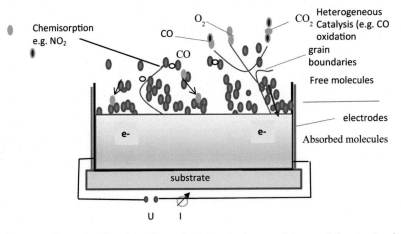

**Figure 9.3** MOS sensor illustrating the detection principles, the layer resistance of the sensing layer as a result of the molecules reaction on the surface (Hurst, 1999). *Source: Redrawn from Hurst, W.J., (1999)* Electronic Noses & Sensory Array Based Systems. *Technomic Publishing Company, ISBN No. 1–56676-780–6.*

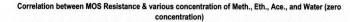

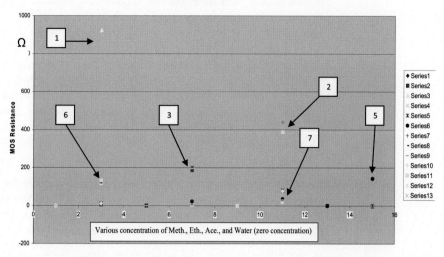

**Graph 9.A3**

versus resistance related positively to each other. Looking at the graph, the following conclusions can be made:

1. Point 1 is the concentration of 10%.
2. Point 2 is the concentration of 1%.
3. Point 3 is the concentration of 1%.
4. Point 4 is the concentration of 0.10%.
5. Point 5 is the concentration of 0.10%.
6. Point 6 is the concentration of 0.05%.
7. Point 7 is the concentration of 0.05%.

All other points below the above points are mainly low concentration or water, or traces from previous experiment still in the chamber.

Thorough cleaning of the chamber was a problem and this was the major obstacle in the accuracy of the test on the MOD. As such, the validity of some of the tests in various MOD systems is affected *if* the above problem *within each MOD system* is not permanently *fully* solved.

The time used for heating each sample ranged from 30 seconds to 16 minutes with changes in temperature from 23°C to 60°C.

It was noticed that the MOD response increased (in the form of the changing of the sensor's resistance) when the temperature for each sample increased, accordingly. Points A, B, and C represent the sensor's resistance on the graph. The changes that occurred were small but still noticeable. It appears that in all the other samples, which were below the first line in the graph, the changes of the temperature did not make much difference. This was either because the concentration was very low (below 0.05%) or because other unknown factors affected the behavior of the sensor, such as the age of the sensor or the environment in the lab at the time of doing that experiment. Some samples displayed little response to the change of temperature, and in other experiments under similar conditions, they displayed much higher responses. These kinds of small but different responses can sometimes be due to a faulty component within MOD itself or due to following the incorrect experimental procedure. In this case, the sensor was replaced with a new one and the same experiments were repeated. However, the results were not that much different from the first one. This means that the first sensor used was not faulty and as such, it is possible that there were other factors causing the different kinds of results. (Graph 9.B3)

The MOD was unused for more than 12 hours in order for the traces of some elements, left from a previous sample(s), to clear from the chamber. In addition to this, earlier on, the time location used on MATLAB software to clean the chamber increased during the end of the last experiment.

On the following day, the same experiments were conducted. However, the results obtained were, again, similar to the previous two sets completed before.

The conclusion from the above, therefore, is either that there was a fault within one part or another in the MOD components or the fault was simply related to the functioning of software itself. (Graph 9.C3)

The different times used to heat different samples had some effect on "some experiments." In this context, using the two words "some experiments" means

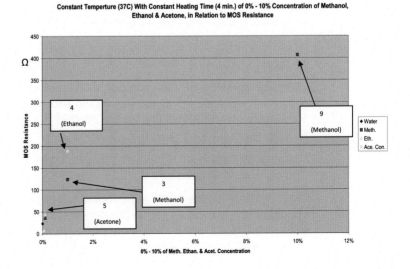

Constant Tempertpure (37C) With Constant Heating Time (4 min.) of 0% - 10% Concentration of Methanol, Ethanol & Acetone, in Relation to MOS Resistance

**Graph 9.B3**

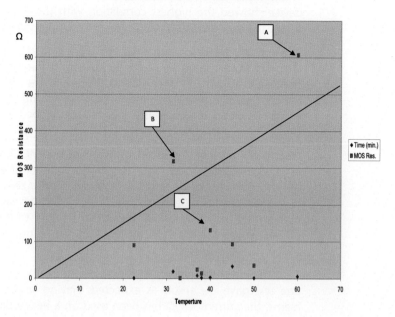

**Graph 9.C3**

that the change of temperature can only have a noticeable effect on higher concentrated samples, that is, the increase of temperature of the water bath played a role in the changing of the sensor response only with samples with high concentrations. There were also changes in the sensor response, using lower concentration in the sample, when the temperature in the water bath increased. However, this change was very small and therefore too insignificant to be considered in such a kind of "basic general test" for the MOD.

## 9.5 Results and discussion

It was found that two of the three different sensors used in the MOD were sensitive for the detection of headspace gas. These working sensors enabled the device to detect various elements in a manner that was most of the time correlated with the concentration of each sample. Thus the present device (built in the laboratory as experimental device) should be able to detect other types of sample concentrations. Furthermore there is a good possibility that MOD system can also detect volatile compounds and organic compounds, such as those emitted from an enclosed biological environment (e.g., growth of bacteria for pharmaceutical and medical use).

In this experimental work, measured concentrations of methanol, acetone, and ethanol, as well as HPLC water, were used in random order with various variables, and many of them were under controlled conditions.

In a comparison with all the samples' results, the measurement modes of the device showed the highest correlation with the highest concentration, which suggested that the MOD was in good working order as expected.

There is the need for improvement related to the way the device cleans its chamber from the traces of the sample from the previous experiment. This is because problems do arise during the test if the device is not cleaned thoroughly before a new experiment. Without solving this issue accurately, errors will accumulate, and obviously, false results will be the outcome.

Another problem was the testing of HPLC water (water vapor). This may be a "traditional problem" noticed with varieties of MODs. What this means is that in using the HPLC water samples during a number of experiments there was an obvious difficulty in keeping the graph line above zero on the Y-axis. A number of suggestions have been made as to the cause or causes of this problem. The outcome of the investigation pointed the finger to the accumulation of vapor on the sensor, the speed of injection and as mentioned above, the cleanliness of the chamber itself.

To eliminate the problem, all the above causes have been dealt with and only one test gave the result as expected. Having said that, despite various problems, the current MOD has been used on a nearly daily basis for more than 3 weeks with most of the results accurate as expected.

### 9.5.1 Other tests

Tests concerning *repeatability, reproducibility, uncertainty of measurement* were performed during the above experiments. However, other tests such as *sensitivity, measuring range, and traceability* were not done, as it should have been done. This was due to problems related to PC and the MOD at the end of this session. Tests such as those related to detection limit and the detection of analyte in a variable sample, as well as procedures and protocols that enable quantitative and objective comparisons to be made between different types of devices, were successfully completed.

To conclude this discussion, it can be said that the MOS response to various samples using the designed MOD experimental model can provide objective measurements, though those obtained with or without minor errors showed a relative and not absolute values.

## 9.6 AI training

To train MOD using specifically a written software program, such as Python either via the use of machine learning (ML), a subset of AI, or via DL layers, a subset of ML. Both ML and DL will be looked at in the following sections.

### 9.6.1 Artificial intelligence

Before going into the training aspect using ML, let us have a brief outlook into AI, in general.

AI is also referred to as "heuristic programming," or "machine intelligence," or "the simulation of cognitive behavior." In few words, AI can be summarized as simulating the way human brain works within a machine.

The above can be achieved via a program that enables the machine to learn and solve problems.

Historically speaking, AI was born out from past philosophical ideas in trying to explain the origin, formation, and progress of our thoughts in the form of processing symbols mechanically. Later on, the above approach prepared the ground for the design of a programmable machine during 1940s implementing a mathematical approach.

During 1955 the logical theory developed by Herbert Simon and Allen Newell is considered by many historians in this field to be the first AI program, that is, by combining data with algorithms, this opened the way for a designed program to learn from the data provided, such as in the form of a pattern occurring and/or certain other recognizable features. The above machine process opened the way to generate a solution for certain problems. In 1956 the term AI was used for the first time at a conference at Dartmouth College, in Hanover, New Hampshire.

AI is used in social media, email, web searching, digital/virtual assistants, agriculture and farming, autonomous flying, retail, shopping, security and surveillance. manufacturing, sports etc. Specific applications of AI in use today are Siri, Alexa, Tesla, Cogito—Boxever, and John Paul. (Box 9.4)

Types of AI are listed under the following headlines:

1. Reactive machines.
2. Limited memory.
3. Theory of mind.
4. Self-awareness.

The most common type of AI is reactive machines.

ML, DL, neural networks are the main three subsets of AI (Fig. 9.4).

When it comes to the domain of AI, this has been listed under the following headlines:

1. Formal tasks.
2. Mundane tasks.
3. Expert tasks.

# Box 9.4 How does artificial intelligence work?

Can machines think?—Alan Turing, 1950

Less than a decade after breaking the Nazi encryption machine Enigma and helping the Allied Forces win World War II, mathematician Alan Turing changed history a second time with a simple question: "Can machines think?" Turing's paper "Computing Machinery and Intelligence" (1950), and its subsequent Turing Test, established the fundamental goal and vision of artificial intelligence. At its core, AI is the branch of computer science that aims to answer Turing's question in the affirmative. It is the endeavor to replicate or simulate human intelligence in machines. The expansive goal of artificial intelligence has given rise to many questions and debates. So much so, that no singular definition of the field is universally accepted. The major limitation in defining AI as simply "building machines that are intelligent" is that it doesn't actually explain *what artificial intelligence is? What makes a machine intelligent?* In their groundbreaking textbook *Artificial Intelligence: A Modern Approach*, authors Stuart Russell and Peter Norvig approach the question by unifying their work around the theme of intelligent agents in machines. With this in mind, AI is "the study of agents that receive percepts from the environment and perform actions." (Russel and Norvig viii). Norvig and Russell go on to explore four different approaches that have historically defined the field of AI: **Thinking humanly, Thinking rationally, Acting humanly, Acting rationally**. The first two ideas concern thought processes and reasoning, while the others deal with behavior. Norvig and Russell focus particularly on rational agents that act to achieve the best outcome, noting "all the skills needed for the Turing Test also allow an agent to act rationally." (Russel and Norvig 4). Patrick Winston, the Ford Professor of artificial intelligence and computer science at MIT, defines AI as "algorithms enabled by constraints, exposed by representations that support models targeted at loops that tie thinking, perception and action together." While these definitions may seem abstract to the average person, they help focus the field as an area of computer science and provide a blueprint for infusing machines and programs with ML and other subsets of artificial intelligence. While addressing a crowd at the Japan AI Experience in 2017, DataRobot CEO Jeremy Achin began his speech by offering the following definition of how AI is used today: "AI is a computer system able to perform tasks that ordinarily require human intelligence. Many of these artificial intelligence systems are powered by machine learning, some of them are powered by deep learning and some of them are powered by very boring things like rules."

*Built in (undated)*

**Figure 9.4**
Deep learning is a subset of machine learning and machine learning is a subset of artificial intelligence.

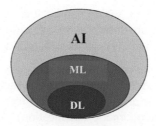

The types of "ML" under AI, can be listed under the following headlines:

1. Supervised (inductive learning)—data include desired outputs.
2. Unsupervised—data do not contain desired outputs.
3. Reinforcement learning—making a sequence of decisions.
4. Semisupervised learning—data contain some desired outputs.

As mentioned previously, Python is the most common language used in AI, which may include the following applications:

1. General AI.
2. ML.
3. Natural language processing.
4. Neural networks.
5. Haskell.

### 9.6.1.1 Machine learning

As has been mentioned in the previous section, ML is a subfield or an application of AI, which focuses mainly on the structure of data and the creation of models. ML is not program instructions but rather a way of training for the computer on the data provided (input data) following statistical methods, that is, in order to deduct the outcome (output data). For the purpose of carrying out the above process, model building from the data provided enables the system to generate the final outcome or a solution.

As has been mentioned previously, ML methods are categorized in different ways, supervised and unsupervised, semisupervised and reinforcement learning. The supervised method is a simple learning approach where an example of input is provided with the correct output, that is, the image of plants labeled as vegetation and image of forest as collection of trees; the learning supervised logarithm, after training, will be able to identify unlabeled image of plants as vegetation and an unlabeled collection of trees as a forest. The supervised is a highly accurate method.

Examples of supervised learning: regression, classification, naive Bayesian model, random forest model, neural networks, support vector machines (SVM).

On the other hand, unsupervised learning is a more complex process in that algorithms look for undetected patterns and information in unlabeled set of input data, which can be less accurate than the supervised method. Examples of unsupervised learning: clustering techniques and self-organizing maps (see Chapter 3: MOD data and data analysis—Section 3.1).

The third method is the semisupervised learning that, as the name indicates, is the middle way between the above two methods in that by providing small number of labeled data in order to support larger and more complex unlabeled data (e.g., image classification).

The fourth method is reinforcement learning—making a sequence of decisions in order to achieve certain goals and obtaining a reward, similar in the way a game is played.

As a reminder, ML is based on some of the statistical principles, such as the relationship between two variables, that is, correlation. While the other statistical term, regression, is in the finding of the relationship, such as strength and character between one dependent and one or more independent variable. Regression is a supervised learning that helps in finding correlation between variables

ML algorithms

Examples of ML algorithms are listed below (Brownlee, 2019):

*Regression algorithms*—Ordinary Least Squares Regression (OLSR), Linear Regression, Logistic Regression, Stepwise Regression, Multivariate Adaptive Regression Splines (MARS), Locally Estimated Scatterplot Smoothing (LOESS).

*Instance-based algorithms*—k-Nearest Neighbor (kNN), Learning Vector Quantization (LVQ), Self-organizing Map (SOM), Locally Weighted Learning (LWL), SVM.

*Regularization algorithms*—Ridge Regression, Least Absolute Shrinkage and Selection Operator (LASSO), Elastic Net (EN), Least-angle Regression (LARS).

*Decision tree algorithms*—Classification and Regression Tree (CART), Iterative Dichotomizer 3 (ID3), C4.5 and C5.0 (different versions of a powerful approach), Chi-squared Automatic Interaction Detection (CHAID), Decision Stump (DS), M5, Conditional Decision Trees (CDT).

*Bayesian algorithms*—Naive Bayes, Gaussian Naive Bayes, Multinomial Naive Bayes, Averaged One-dependence Estimators (AODE), Bayesian Belief Network (BBN), Bayesian Network (BN)

*Clustering algorithms*—k-Means, k-Medians, Expectation Maximization (EM), Hierarchical Clustering.

*Association rule learning algorithms*—Apriori algorithm, Eclat algorithm.

*Artificial neural network algorithms*—Perceptron, Multilayer Perceptrons (MLP), Backpropagation, Stochastic Gradient Descent, Hopfield Network, Radial Basis Function Network (RBFN).

*DL algorithms*—Convolutional Neural Network (CNN), Recurrent Neural Networks (RNNs), Long Short-term Memory Networks (LSTMs), Stacked Auto-encoders, Deep Boltzmann Machine (DBM), Deep Belief Networks (DBN).

*Dimensionality reduction algorithms*—Principal Component Analysis (PCA), Principal Component Regression (PCR), Partial Least Squares Regression (PLSR), Sammon Mapping, Multidimensional Scaling (MDS), Projection Pursuit, Linear Discriminant Analysis (LDA), Mixture Discriminant Analysis (MDA), Quadratic Discriminant Analysis (QDA), Flexible Discriminant Analysis (FDA).

*Ensemble algorithms*—Boosting, Bootstrapped Aggregation (Bagging), AdaBoost, Weighted Average (Blending), Stacked Generalization (Stacking), Gradient Boosting Machines (GBM), Gradient Boosted Regression Trees (GBRT), Random Forest.

ML is one of the methods that can be used successfully to train MOD devices for more accurate outcomes; however, there can be limitation compared with DL as the more the device is used via DL, the more algorithmic learning takes place within the device AI.

### 9.6.1.2 Deep learning

This is an approach designed to mimic the function of the brain via the utilization of neural network, where a number of processing layers between the first (input) and the last layer (output) are located (hierarchical representations) for the purpose of each layer to identify certain aspects of an image, speech, or information provided. These are layers of nonlinear processing units, that is, multilayer structure of algorithms, which are referred to as neural networks. These layers work via examining or selecting basic elements or parts (such as of an image). These elements are the inputs for the following layer, which in turn sends the output to the next layer and the process continues from one layer to another. The above means that computation takes place within each layer.

Some of the popular applications for DL are image or video processing (convolutional neural network), natural language processing (RNN), and audio data (standard neural network). Also for other various applications, including MOD devices, DL can be used successfully as well.

A question may arise regarding the difference between DL and neural networks? Neural networks are where the DL takes place, but the process is called DL when there are many hidden layers within the neural networks itself (Fig. 9.5).

DL is therefore, in general, mimicking what takes place during various processes in nature but particularly the way how species learn.

The question is why not just implement ML instead of DL for the MOD? Of course either of the above methods can be applied and worked for the device;

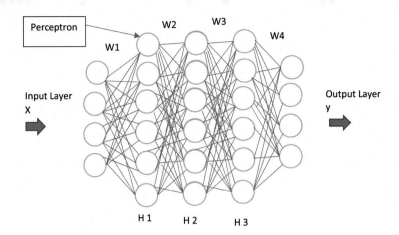

**Figure 9.5**
Schematic
illustration of
neural network
layers.

however, complex analysis of odor in certain situations with larger amount of data (where ML may not be able to function correctly, i.e., saturated) can only be done using DL for the result required with less human interference.

### 9.6.1.2.1 Deep learning framework

For designing, validating, and training, building blocks are needed, which DL framework can provide. DL framework is described as a tool (interface) that helps in the construction of models via the use of its various components.

The following are selections of DL framework:

TensorFlow (a popular tool for ML and DL.), PyTorch, Keras, Sonnet, MXNet, Swift for TensorFlow, Gluon, and DL4J.

### 9.6.1.2.2 Activation functions

Data that can be expressed numerically, in the form of a small value, received via an input node where mathematical equations, that is, activation functions, determine the outputs for the neural network. These activation functions are part of each neuron and decide on activation (firing) or not, that is, it is like checking if the information a neuron is receiving is relevant for prediction or should be ignored. The higher the number the greater the activation.

The activation function is a way of checking an incoming value. The final outcome will be decided via the output nodes before being released.

The general advice on which activation function should be implemented can be briefly outlined as follows:

Reportedly, the issue connected to the vanishing gradient means that Sigmoid (Sigmoid Function) and tanh (Hyperbolic Tangent Function) are not suitable to use. On the other hand, Relu (Rectified Linear Unit) proved to be a good choice during implementation for hidden layers (generally speaking, most convolutional neural networks or DL use it). Softplus and Softsign are not suitable either, while for deep networks (DL) it has been reported that Swish (from Google Brain Team) can be a better option than the use of Relu.

### 9.6.1.2.3 Bias

The bias play an important role as it adjusts the output together with the input value and neuron weight such as y (output) = $\Sigma$(weights $\times$ inputs) + bias.

Bias is an additional input into the next layer that has the value of 1, that is a node which is always switched on, if we can use this term. Without a bias, the gradient graph will always pass through 0, which means it will not be able to generate outputs differing from zero in the next layer.

> Without a bias neuron, each neuron takes the input and multiplies it by a weight, with nothing else added to the equation. So, for example, it is not possible to input a value of 0 and output 2. In many cases, it is necessary to move the entire activation function to the left or right to generate the required output values—this is made possible by the bias. *(MissingLink.ai, undated)*

So bias works by adding a number, that is, a constant that helps the model best fitted for a given data, such as to shift the activation function to the right or to the left, that is, similar to an intercept added in a linear equation. Bias acts as another specific neuron as well to assess the model prior to training (Fig. 9.6).

## 9.7 Machine olfaction device and artificial intelligence

In this final part of the chapter, consideration will be given for the outline design of the Python program prior to the programming, but not necessarily

**Figure 9.6**
Schematic illustration of neuron, bias, and activation function process.

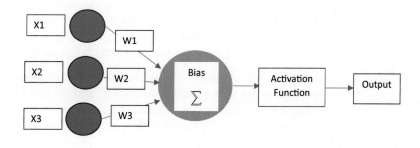

about the coding itself. In order to do that, the first thing to be taken into consideration is what kind of detection or application (or applications) our MOD should be designed for. To illustrate, let us consider two options:

1. One application.
2. More than one application.

Whether one application or more, the need for the requirements of the user should be considered in advance. For example, in the case of one application, such as within the medical field, for detecting specific illness, such as cancer, focusing on one particular area for example, lung cancer, the output, therefore is simply to find out whether a patient has cancer in the lung or not. If, on the other hand, the MOD device is used in many different applications, then DL layers have to be trained on a variety of cases and obviously the length of time related to this kind of training should be taken into consideration as well as the access to the variety of data required.

## 9.7.1 Software development plan

There are a number of factors that should be taken into consideration for the development of a software program (Gao & Feng, 2012), such as for the Python program for our intended MOD:

1. Type and size of the software—It is important to consider the differences when it comes to the size and type of the program, for example, architecture and amount of data needed are different.
2. Experience in this field from predecessors—Learning from the development of present and past similar software.
3. Difficulty or the type of difficulty in obtaining user demand—How to obtain real and detailed user requirements.
4. Development of techniques and tools—The techniques and tools that are required for the development in helping to achieve the aim of the software.
5. Development team and related background—The team developing a particular software application should have good background knowledge in the application field.
6. Risk during development of the software—Awareness of possible risk during each stage of the software designers' development plan.
7. Which software development methods will be needed? —There are many software development methods; therefore it is important to consider carefully which will support the aim of the application.

Assuming the coding of the Python software program is fully completed, the question arises about testing the system and related procedures and time allocated for this part of the project. The software engineer will test the program relevant to the application he/she compiled. The above obviously will not be enough unless of course the actual user (customer) of the device will be able to run the program successfully with accurate repeatable results.

If a Software Requirements Specification has been developed, it describes all user-visible attributes of the software and contains a Validation Criteria section that forms the basis for a validation-testing approach *(Pressman & Maxim, 2015).*

# References

Brownlee J. (2019) *A tour of machine learning algorithms. Machine Learning Mastery* <https://machinelearningmastery.com/a-tour-of-machine-learning-algorithms/>

Built in (undated) *Artificial intelligence. What is artificial intelligence? How does AI work?* <https://builtin.com/artificial-intelligence>

ChemicalSafteyFacts.org (undated) *Acetone (Propanone)* <https://www.chemicalsafetyfacts.org/acetone/>

Gao, Y., & Feng, X. (2012). Factors to be considered when to design software development plan. *Advanced Engineering ForumOnline, 6 − 7*, 3−8, 2012−09−26ISSN: 2234−991X.

Hurst, W. J. (1999). *Electronic Noses & Sensory Array Based Systems.* Technomic Publishing Company, ISBN No. 1−56676-780−6.

MissingLink.ai (undated) *Neural network bias: Bias neuron, overfitting and underfitting.* Neural Network Concepts <https://missinglink.ai/guides/neural-network-concepts/neural-network-bias-bias-neuron-overfitting-underfitting/>

Pressman, R., & Maxim, B. (2015). *Software engineering: A practitioner's approach* ((8th ed.)). Mc Graw Hill.

PubChem (undated) *Methanol (Compound summary).* National Centre for Biotechnology information. National Library of Medicine. <https://pubchem.ncbi.nlm.nih.gov/compound/Methanol>

Strathamann, S. (2001). Sample conditioning for multi-sensor systems. <https://publikationen.uni-tuebingen.de/xmlui/bitstream/handle/10900/48144/pdf/StraThesis.pdf?sequence=1>

World of Molecules (undated) *Chemical and physical properties of the ethanol molecule* <https://www.worldofmolecules.com/fuels/ethanol.htm>

# Review and MOD future implications

In this final chapter of the book, a general examination, review, the application marketplace, and future predictions for the machine olfaction device (MOD) are presented. This part of the book can be one of the ways to help in stimulating further reader thoughts about an important field of work and research, such as within the field of MODs.

Alexander Graham Bell thought about the discovery of a new science in detecting odor (see Chapter 1: Background, Materials and Process—Section 1.1.1). This new science, as referred to by Bell, is already a fact from both the technical and the scientific approaches.

Many of the challenges MOD initially faced, such as short lifetime of the sensors, sensitivity toward moisture, narrow selectivity, relatively expensive, mostly used as a laboratory tool (large and nonportable) with results that could be difficult to reproduce, were due to the presence of multiple sensors. These shortcomings are mostly already solved, specifically the technical parts, via a new design implantation such as those areas within the field of microsystem, which are already being implemented commercially, or even aspects connected to the implementation of nanoelectronics, if and when these commercially take place on a large scale.

The following therefore are summary solutions for the above issues as provided in the new MOD design, presented in this book.

*The short lifetime of the sensors*: The design of the chambers provided in this book will solve the problem, regardless of whether the life cycle of the sensor is short or long!

*The sensitivity toward moisture*: The new chamber design will prevent any amount of moisture forming.

*The narrow selectivity of the device*: This issue can be managed via the training provided, using deep learning.

**Introduction to Machine Olfaction Devices.**
DOI: https://doi.org/10.1016/B978-0-12-822420-5.00006-4

*Drift*: Calibration (regular calibration to the sensor done automatically by the device) or replacing the old sensor with a new one, as the device will inform the user when replacement is needed. However, there is the need for standard calibration, if and when there is commercial production for the MOD on a large scale. The calibration can be done online as either the device communicates directly with the standard source (e.g., manufacturer, software company, or organization) without the need for the user action or it can be done manually by connecting the MOD with a PC for an update in this respect.

*The price of MOD*: The price is certainly declining at the present time and will do so even further in the near future as the technology improves, which, together with wider global investment in this field, will help in reducing the cost. As progress is presently taking place in printed electronics, it will certainly have direct impacts on the prices of electronic devices in general and that includes major reductions in the prices of the MOD as well.

*MOD results can be difficult to reproduce*: The number of identical sensors used in MOD may not produce exactly the same response when exposed to an odor under the same conditions. One possibility for this can be originated from the manufacturing process. In literature, this is referred to as sensor discreteness. Via the use of artificial intelligence (AI), together with latest advancements in sensor technology, such as the uses of a quartz crystal microbalance (QCM) and single nanoparticle-based surface plasmon resonance, in particular the implementation of the method discussed in Chapter 5: Biosensors and MOD Design (Section 5.2.1), the repeatability will be close to 99%, if these are implemented together.

## 10.1 Objectives and review

What are the objectives the new MOD design requires to achieve and if they have been achieved or not? The objectives are as follows:

1. Faster stabilizing base line.
2. Quicker sample result display.
3. Ability to use ambient air.
4. Low power consumption.
5. The ability to deal with different varieties of organic/inorganic samples.

*Faster stabilizing base line*: The selection of the sensors, including low-cost common sensors, such as TGS2600, TGS2602, TGS2620, TGS2611, is an important first stage in minimizing the above phenomena. Metal oxide (e.g., $S_nO_2$) sensors suffer from the phenomena of drift, which means either regular replacement will be needed, or the software controlling the working mechanism of all the sensors should be considered. What this means is that, whenever there is drift from a particular sensor, which can be detected using ANN, the reading will be directed to another two sensors for comparison and only after ANN selects the correct reading, usually generated by the two sensors and confirmed via ANN after comparison between the two sensors and the original drifting sensor, then the result will be released.

*Quicker sample result display*: The type of processor used and the quality of the written software, together with a highly sensitive/high performance sensor as well as the type of application used within the device, and the condition of the environment where the device is located, all have certain influence, directly or indirectly, on the displayed result. More importantly, the design of the device, whether in relation to the hardware and/or the software, also plays a major role in this respect.

*Ability to use ambient air*: This depends on the application the device is programmed to perform. In many cases, the difficult part is when the device is used in field applications, where changes in environmental conditions may have impacts on the sensors. However, as these conditions, such as temperature, pressure, humidity are adjusted accordingly within the internal environment of the device, as well as the way the chambers have been designed to minimize or prevent this kind of impact on the sensor, plus the type of sensor used. Metal oxide semiconductor (MOS) may not be the right sensor as it is not possible to discriminate the interference gases from the target gases and therefore a different type of sensor(s) should be employed. In addition it depends on the number and types of interference gases, if and when the device is used within enclosed environment, as the open environment may have vast number of interferences. Taking the above into consideration, it will enable the MOD to be used in sampling ambient air. Specific requirements, such as using one type of application only will, however, improve the outcome.

*Low power consumption*: Similar to the technology presently used in minimizing the usage of power in a mobile phone, this facility is implemented in the mobile MOD device. However, there are many other factors that can lower the usage of power in the new MOD design. The implementation of nanoelectronics, such as nanotube sensors, will reduce the power consumption drastically. The type of battery used is also another factor, which needs to be taken into consideration (see Chapter 5: Biosensors and MOD Design, Section 5.2.1).

Board circuit design, such as the use of integrated circuit or microsensors based on silicon and the materials used, provides a low power consumption as well as lowering the overall cost. The new MOD design includes some of the above. However, keeping in mind that mobile MODs (or the hand-held MODs) their functions are greatly reduced compared with bench-top MODs, due to the changing environment and constraint on the power supply (Box 10.1).

---

## Box 10.1 Constraints

Since there are severe constraints on overall size, weight, and power consumption, the performance of hand-held devices is generally inferior to that of bench-top instruments. Precision and the limit of detection (LOD) are degraded by a factor of approximately 10 as compared to a bench-top setup due to the less effective temperature stabilization and due to the use of less bulky measurement and recording electronics such as no precision voltage sources, less-sophisticated counter modules, and off-the-shelf electronic parts. The hand-held units are less pricy and intentionally designed to suit only a few specific applications, in consequence their versatility is rather limited. The unit configuration usually has to be optimized with regard to the target application.

(Hierlemann, Weimar, & Baltes, 2002).

---

*The ability to deal with different varieties of organic/inorganic samples:* The question is whether the device is designed for one application or for different applications? Different applications will enable the MOD to deal with varieties of organic/inorganic samples. What this means is that within the device, each sensor is specialized in detecting one to four different types of odors. As the device contains an array of sensors, by switching between one sensor to another, using AI deep learning, the device can detect a particular odor searched by the user. This of course depends on the number of odors the MOD is required to detect, as well as the number of sensors installed within the device.

## 10.2 Review of nanotechnology

Implementing nanotechnology within various industries is vital, specifically when introduced to devices' electronic circuitry, as an important part of the manufacturing process for speed, accuracy, and cost reduction. Nanotechnology benefits cover many fields and applications, such as within the field of nanoparticles, for coatings (see Chapter 8: MOD Nanostructure Coating), as well as in chemical processing and helping in drug delivery, nanomaterials for improved materials in manufacturing industries, nanoelectronics, for a variety of electronic devices, nanooptics at nanoscale making of optical circuits, nanobioelectronics, a hybrid of nanodevices and biological materials, nanomagnetics, nanomagnetic materials and devices for magnetic storage and logic devices and nanofluidics, designing devices by employing fluid properties at nanoscale level.

In the case of the MOD, the nanoscale internal structure design is what the future is all about. For this reason, nanoelectronics will and should play a major role in the development of a variety of manufactured products, sooner than later.

(The book discussed metallurgical work in the form of sampling and analyzing, as well as nanotechnology tools, such as microscopy and related areas using various experiments and samples, plus the examination and analysis of the nanostructure of coating for the MOD.)

Simply, the nanotechnology approach is the manipulation of atoms, molecules, and macromolecular structures, as well as exploiting the unique properties of matter at a nanoscale (1–100 nm) level, a level that describes this kind of human engineering activity inside matter. Here we are dealing with nanostructures—a nanometer is one-billionth of a meter, about three to four atoms in width. That is a thousand million times smaller than a meter. To give you an example how small this scale is, just imagine we enlarge an orange to the size of the earth, then the atoms of the orange would become visible, and the size of each atom will be about the size of a single grape.

Nanotechnology is the result of a marriage between chemistry and engineering. Therefore it is like traditional chemistry but without chemical reaction. Here we are simply building things using one atom or molecule at a time via programmed nanoscopic robotic arms, or similar methods. The question, therefore, is that is it possible to move one atom (or one molecule) from one location and locate it to the *desired* place without any difficulty—simply using the technology available to us now?

The answer of course is yes, but with some difficulties. At the same time, this process is somehow limited at the present time, and therefore transferring large number of atoms/molecules in a much faster way and on a large industrial scale may not be viable with today's present technology.

The emphasis is that nanotechnology is a field of engineering. Therefore to understand these processes, using more sophisticated technology, it is possible to say that by treating atoms or molecules discretely in a way similar to the process of computers treating bits of information, we can build anything we desire from within—at a very little cost. This is of course the theory, the practical part of how to apply it, similar in applying any technical or engineering method used today for various products, is still in the realm of the future. But suppose we can apply this technology today on a large scale. Then this could mean an automatic construction of consumer goods without traditional human labor, as we know it. Obviously any number of copies can be produced at no extra cost whatsoever. This is where self-assembly comes into the picture. A good example of self-assembly is taking place in nature itself around us, that is, an atom in a cell in a vegetable manipulating the atoms of soil, air, and water in order to produce more copies for growth to continue—or arranging the atoms in coal— the results are diamonds. Therefore manufacturing, using the principles of nanotechnology, which is happening on a limited scale today is expected to undergo a profound change in the future. Advances in miniaturization will reach the level of individual atoms and products will be designed and built to atomically precise specifications. Though a large part of nanotechnology is still within the domain of laboratories and limited products, that is, within a limited scale within the manufacturing industries, increasing governmental and private funding in various research in this field will pay off eventually. Nanotechnology will be fully an essential part in the near future in relation to the design and manufacturing of various electronic devices, including MOD (Box 10.2).

---

## Box 10.2 Nanotechnology and electronic noses

Nanotechnology is seen as a key in advancing e-nose devices to a level that will match the olfactory systems developed by nature. Nanowire chemiresistors are seen as critical elements in the future miniaturization of e-noses. It is now also believed that single crystal nanowires are most stable sensing elements that will result in extending of lifetime of sensors and therefore the recalibration cycle. Last year we reported on a research effort "Towards the nanoscopic electronic nose." Scientists involved in this effort now report a second-generation, far more advanced e-nose system based on metal oxide nanowires.

"Despite encouraging demonstrations of an array of individual metal oxide nanowires, there still exists a technological gap between the laboratory demonstrations and a practical e-nose microdevice suitable for up-to-date large-scale microfabrication and capable of operating in real-world environments," Dr. Andrei Kolmakov explains to Nanowerk. "Hence, our aim was to bridge this gap and demonstrate the excellent performance of a practical device made by combining 'bottom-up' fabricated $SnO_2$ nanowires/nanobelts as sensing elements with a 'top-down' technology of the state-of-the-art multielectrode KAMINA platform."

*Nanowerk*

(Berger, 2007).

---

To enable us to understand various basic tools and mechanism—which form an important part of nanotechnology—the subject of microscopy and its related tools have been already presented in Chapter 7: Nanoelectronic Systems. The following is a brief review.

The idea of producing a microscope, such as the scanning probe microscope (SPM), started from the early years of the last century; however real results only began to be collected during the early 1980s. The first scanning tunneling microscope (STM) was invented by G. Binnig and H. Rohrer in 1981 (IBM, Zürich), In 1986 G. Binnig, C. F. Quate and Ch. Gerber presented another SPM, called the atomic force microscope (AFM).

How does it work? Well, when we bring a probe (tip) very close to a surface (near 1 nm), different physical phenomena appear, which, if we scan the surface very precisely, allow us to obtain information at the nanometric level. For that reason, the term scanning probe is employed.

Electrons are transferred between the surface and the tip if we polarize a conducting or semiconducting sample with a conducting tip.

The tunneling effect is the name of the quantum phenomenon, which produces this current. When scanning the surface with a piezoelectric ceramic attached to the tip, which allows very small displacements, we record the variation in tunneling current, which is proportional to the tip-surface distance. This technique is called scanning tunneling microscopy (STM). In the case of very planar surfaces, it is possible to actually "see" atoms and even manipulate them.

Various positional devices—especially large ones, which we use in SPMs—have been especially designed to be stiff in order to image individual atoms, despite the problem of thermal noise that exist in this kind of environment.

The recorded images are digital, generally as $512 \times 512$ point plots. Each point has a value between 0 and 255 so that the greater the value, the more elevated the point and vice-versa. Therefore in this way we can represent 3D information in 2D. When we present a 2D image, we also include the Z range to determine the height difference between black and white areas.

Atomic force microscopy uses the same tools of tip and surface. In this case, a tip attached to a cantilever of fixed ring constant approaches the surface. An interaction involving different types of forces (Van der Waals' force, capillary force, friction, etc.) appears between them. In measuring the force variation during scanning, we record the height variations on the surface. This technique has the advantage of being able to function under widely varying environments (ambient, liquid, electrochemical etc.) and on either conducting or insulating samples.

In SPM and as have been mentioned above, a very sharp tip is brought down to the surface of the sample being scanned. It is possible to tell that the tip is approaching the surface and so the outlines of the surface can be sensed. Various ways are used to detect the surface; this means some SPMs touch the hardness of the surface. Different types may get connected to the surface and measure the current flow between the surface and the source. There are other types, but the principle is still the same, which means when the sharp tip of the probe approaches the surface a signal is generated. That signal will let us map out the surface we are examining.

With today's SPM capability, chemists have already achieved a large size of structure that can be built with no error. In this case, each molecule had 173 atoms, that means over 1000 atoms. SPMs can see the structure that is being built and provide constant feedback to the operator, which helps him/her to detect and correct errors. As you can see, we are still far behind when it comes to the actual work and speed of SPMs as their work should be done in a fraction of a second rather than what is happening today, as it takes a hours to arrange a few molecules or atoms.

Using an AFM, a second approach for positioning molecular building blocks in solution can be achieved. This type of SPM relies on touch to create an image. By pushing on the structure being scanned, and feeling how hard it pushes back, the AFM can build up an image of stiff structures (it does not work very well when the applied force is strong enough to deform the structure). Because the AFM touches the surface of the structure being probed, it can also change that surface. To do this, it is very useful if the precise molecular structure of the tip can be controlled, so that the precise nature of the tip-surface interaction can be well-defined.

Whether through self-assembly, by improvements in SPMs, some hybrid approach, or perhaps by some other path; we are moving from an era of expensive and imprecise products to an era of inexpensive products of molecular precision.

## 10.2.1 Robotic arm

The goal of early nanotechnology is to produce the first nanosized robot arms capable of manipulating atoms and molecules into a useful product or copies of itself—on a commercial scale.

The fact is that it will take a long time and continuous research before beneficial results for the general public—in this field—can become part of our daily reality. It took over two decades and after extensive nanotechnology research got underway for the creation of the first working nanotech robotic arm.

This, therefore, means if anything that nanotechnology, as a practical technology and like any present-day engineering, is already gradually establishing itself.

To arrange molecules or atoms in the right place, as we require to do in nanotechnology, Richard Fevnman the Nobel prize winner physicist said in 1959 *Nothing in the laws of physics prevent us from arranging atoms the way we want . . . it is something in principle, that can be done.*

If we are to position molecular parts, we must develop the molecular equivalent of "arms," that is, the use of robotic arms, as some writers call them, and know how to pick up two molecular structures and put them together. The measurement of robotic arm is 100 nm high and 30 nm in diameter; it has few million atoms and about a hundred moving parts.

The Stewart platform device, which has six adjustable edges (in compression or tension), has the advantage that it is much stiffer and simpler than a robotic arm. Diamond and graphite both are stiff and has remarkable properties, which can be used for this purpose.

It is even better if hydrogen or carbon molecule is added to diamond (flat diamond surface and add two carbon atoms to it) for a stiffer arm.

## 10.2.2 Self-assembly

We know that the basic principle in self-assembly—in molecular structures—is selective stickiness. That means if two molecular parts have complementary shapes and charge patterns, that is, one part has a hollow where the other part has a bump, and one part has a positive charge where the other part has a negative charge, then they will stick together in one particular way. This principle can be used in nanotechnology to assemble what we want.

Some scientists believe that self-assembly is an important part of nanotechnology, but the technology for it has not arrived yet, even though we can see that process happening everywhere in nature.

Positional control plus appropriate molecular tools should let us build a truly staggering range of molecular structures; however, it is too costly. Having said that, building range of molecular structures will happen when someone designs a general-purpose programmable manufacturing device that is capable to make copies of itself, then the manufacturing costs for both the devices and anything they make will be low. A device like this one has been named by Drexler as Assembler. In 1940 Von Neumann produced the first analysis related to self-replicating systems. The architecture for Drexler's assembler is a specialization of the more general design proposed by Von Neumann. Other examples of self-replicating systems are Internet Worm, *Mycoplasma genitalia*, human genes etc.

What we have seen so far is that there is a desired goal for nanotechnology to produce a system able to inexpensively synthesize most diamondoid, that is, strong—and other type of structures and materials. Therefore the intention for nanotechnology at the present is a general-purpose programmable manufacturing system, which uses positionally controlled highly reactive tools in vacuum and is able to self-replicate.

The question remains—is it possible to design a system able to build an assembler with hundreds of millions or billions of atoms with no atom out of place? If we can do that, then the error rate must be low or if this is not possible then we must have an error detection and correction system in operation at the same time.

At present SPMs can build a small part of the size of possible assemblers and have error rates high that will need to use an error detection and correction methods alongside it. The best possible solution is to build assemblers in which error rates are very low and therefore there is no need for an additional system for error detections.

Another problem scientists have discovered is the difficulty to build an assembler within a vacuum and therefore various solutions were put forward to overcome this problem. One of these methods is called building block-based nanotechnology. That is building other materials from large molecular building blocks in large numbers to reduce the number of assembly steps. The whole process is done within a soluble solution to overcome the need for a vacuum.

Solution-based systems could use positional control to assemble the building blocks, but can also use the methods of self-assembly.

## 10.2.3 Nanomachines

Nanotechnology uses well-known physical properties of atoms and molecules to make novel devices with extraordinary properties. Nature uses molecular machines to create life. Scientists from several fields such as chemistry, biology, physics, and electronics are trying to manipulate matter on the atomic scale. Research programs in chemistry, molecular biology, and scanning probe microscopy are laying the foundations for a technology of molecular machine-systems, what we call nano-machines. The structure is similar to an average machine with its power structures and generator except they are built using atoms/molecules to construct their parts. *A general-purpose molecular assembler-arm must be able to move its hand by many atomic diameters, position it with fractional—atomic—diameter accuracy, and then execute finely controlled motions to transfer one or a few atoms in a guided chemical reaction* (Dr. K. Eric Drexler, Institute of Molecular Manufacturing).

A protein is a molecular machine that routinely manipulates individual atoms. Proteins have physical structure and functionality; protein engineers can now synthesize all 20 common amino acids, which are the building blocks of proteins. They have even been able to design synthetic proteins with novel properties. Scientists are racing to catalog the functions of proteins, how they fold, and discover properties of synthetic proteins. Chemists are synthesizing larger and more complicated molecules that perform complex physical tasks. Chemists and biologists produce ever more complicated self-assembling molecular structures.

To build nanomachines, we can say that molecular bearings will be a fundamental component of many molecular mechanical devices. Past work on micro-machine bearings has produced poor results, both experimentally and in theory. There is no use in lubricating the bearing in nanomachines, for the viscosity of the lubricant effectively increases as the scale is reduced.

A molecule that is held together by a single *sigma* bond and does not have any interference between the two parts of the molecule will allow one part to rotate freely with respect to the other. That bond can serve as a bearing with load strength of the order of nanonewtons.

Some bacteria can swim, and while advanced cells can swim using *flagella*, bacteria have simple helical rods of protein that rotate. When the rod attaches to the cell wall there is a variable speed reversible motor. This motor has been described in the literature as a "proton" turbine.

Does present technology succeed in making a nanomachine or a nanomotor? As far as 2003, Technology Research News (March 11) reported from an article that appeared first in the February issue of the *Journal of Biological Chemistry* that researchers from Purdue University have constructed a tiny motor from

DNA and RNA molecules. The device, fueled by ATP, the molecule that powers our own movements, could eventually power nanomachines. The motor measures about 30 nm long, which is less than one-hundredth the size of a red blood cell. It is made from six strands of RNA surrounding a central strand of DNA. In the presence of ATP, the RNA strands push the DNA axle in succession, spinning it around. This produces $50 - 60$ pN, or trillionths of a newton of force. A falling apple exerts about 1 N of force.

This motor has potential in biological applications as well. The researchers have driven the tiny motor axle through the protective protein coat of a virus. The motor could be used to deliver genes or therapeutic molecules into live cells.

## 10.3 Device sampling

*Sample lifetime:* Depending on the type of samples required, certain samples, in particular organic samples, may undergo changes if not kept under optimum conditions. Changes in temperature, sunlight exposure, type and condition of the sample container etc. may impact on the sample in various way, including the starting of a chemical reaction or even the disintegration of the sample. Collecting a sample and the method of preparation for a sample(s) should be carefully considered prior to the detection needed via MOD (Box 10.3).

---

### Box 10.3 Reliable automated airflow control design

The airflow control design is a critical element in the e-nose system, analogous to the important and necessary activities of air inhalations that control the airflow in the human olfactory system. An e-nose's air intake system contains the mixing chamber, the gas chamber, the driver circuits, and the actuators (the sampling pumps and the solenoid valves). The primary function of the air intake system is to ensure the odor tests are executed routinely and automatically by controlling the actuators.

(Zang et al., 2020).

---

*Tubing system*: An important consideration should be made when selecting the tube that delivers the odor (volatile compounds) to the sensor in that the materials of the tube should have no impact (neutral) on the sample odor, that is, no chemical reaction, modification, or being ab/adsorption via the wall material(s) of the tube. The passage of the odor may not be only via the delivery system, which means the chamber and other areas of the device hardware should be considered as well (Box 10.1, Box 10.4).

---

## Box 10.4 Delivery system

The delivery system is composed of two air pumps, the sample chamber, and the relays. The sample chamber is in line with one of the pumps, being called the exposure pump, before connecting to the detection chamber. This allows the air containing VOCs in the sample chamber to be carried into the detection chamber. The other pump, the recovery pump, connects directly to the detection chamber as its purpose is to restore the initial conditions in this last. Both pumps use ambient air. In a hierarchical order, the control system informs the relays when and which pump should they power on, then the pump either carries the VOCs to the detection chamber or renews the air inside it, depending on which pump was powered. By switching the activation of the pumps, the VOC exposure/recovery cycles are generated.

Santos et al., 2019.

---

*Setting up the temperature for the sample incubation:* Increasing or decreasing the sample temperature may have an impact on substance volatility and either increasing or decreasing the concentration in the headspace. The length of the time in delivering the sample or the time period of a stored sample before being used may cause a change in temperature of the sample, which can be the cause of contamination, such as in the form of bacteria and other types of microorganisms, specifically if not stored under specific conditions.

The increasing temperature of the sample as well as a lengthy equilibrium period will have impacts on the odor volatility and may increase headspace concentration (Silvello and Alcarde, 2020).

Sample preparation/size of the sample: Correct sample preparation/handling and the size of the sample is an important part of minimizing error or inaccurate results. According to Burlachenko, Kruglenko, Snopok, and Persaud (2016), a number of factors should be considered during sample preparation:

1. Sufficient number of components in the mixtures should be studied.
2. Low concentration of targeted element, that is, specific to the character of the provided product.
3. What is the dominant component, such as water, in the sample to be examined?
4. Knowing the sensor cross-sensitivities, that is, similar outcome to different substances.
5. The concentration/aggregation state of the sample.

### 10.3.1 The rate of sample injection and the quantity of injection

The rate of sample injection may vary depending on the type of sample, headspace, or concentration of the substance we are looking for in the sample and whether the MOD is used in closed environment or open environment. There are other factors as well that can be related to the application the MOD is used for. The above can be applied to both, that is, the rate of injection and quantity of injection.

### 10.3.2 Sample flow method

The sample flow method or odur delivery system can be achieved by implementing one of the many commons methods such as headspace sampling method, bubbler, sampling bag, diffusion, and permeation method. The above methods are briefly described below (Pearce, Schiffman, & Gardner, 2002).

A. *Head space sampling method*: This is a simple and easy method to implement. Using the space above the sample (usually a liquid sample), occupied by the vapor/gas, the vapor/gas is carried to the sensor(s) via the carrier gas. The flow is controlled by a mass flow controller. It should be noted that the point where the vapor/gas (e.g., the level of the needle above the liquid) should stay at the same level while taking the sample.

B. *Diffusion method*: This is another simple way in taking vapor/gas sample via diffusion. However, as the concentration of the vapor may not be that accurate, there is a possibility of error. A tube with its dimension has been measured, with low concentration of the vapor/gas is measured. As the liquid evaporates it will slowly diffuse via a diffusion tube and pass to the sensor at a controlled rate.

C. *Permeation method*: There is a similarity between this method and the diffusion method in that both require the same design except that the diffusion here takes place via a tube. By passing a liquid, a gas/vapor from the liquid will pass through the tube walls. As in any other sampling flow method, the temperature should be controlled. The length of the tube is another important factor in this method.

D. *Bubbler method*: Another simple method in obtaining vapor sample via the use of a container for the liquid sample. By inserting air via a tube inside the liquid, bubbling will take place and the generated vapor is taken away as a sample. This method may not be a good method as in many cases the headspace does not saturate sufficiently in order to obtain the required sample.

E. *Sampling bag*: By injecting large air volume sampling bag with the intended liquid sample to be examined, the vapor generated from the sample liquid is sucked out of the bag and passed to the sensor. If more than one sample is needed for different elements, then a mass flow controller will be used for this purpose.

F. *Static system*: By having a vapor at constant concentration at a constant temperature, it will enable the sensor, after equilibrium is reached, to analyze the sample. This takes place when the sample is provided at a constant rate, which usually happens via the use of this system.

## 10.4 Sensors

The present development of MODs compared to what it was a decade ago showed that a major advancement has already taken place, whether related to the advancement in sensor design and manufacturing, or within the field of micro and nanoelectronics. AI as part of the recognition system plus related software, applications accuracy, the size of the device, and the beginning of lower prices for a variety of MOD devices.

However, sensors and the quality of sensors in relation to specificity, selectivity, repeat accuracy, resolution, accuracy, sensitivity, and life cycle, all play a major role in the outcome of the device (Table 10.1). Therefore sensors are the most important part of the MOD system. As the new MOD design in this book uses two different types of sensors, that is, biosensors and technical sensors, a comparison between the above two sensors is useful to be aware of in order to help in choosing the right sensor(s) for the required function or application. According to Full, Delbrück, Sauer, and Miehe (2020), for the above-mentioned criteria, it has been found that the biosensors achieved better than technical sensors, except for repeat accuracy. The authors also mentioned that technical sensors perform better than biosensors in areas such as durability, maintenance effort, and resistance to environmental influences.

What is lacking with the present MOD technology/sensors is that the concentration of a particular element or odor in a sample that the device does not provide. This is because during MOD training, the concentration of an odor in the sample should be known; however, after MOD training, the device is able to provide an estimate of the odor concentration if and when certain correlation takes place during the training between the dynamic olfactory results (see Chapter 1: Background, Materials and Process, Section 1.7,) and the device sensor response (Bax, Sironi, & Capelli, 2020).

Some of the sensors, such as such as surface acoustic wave sensor, QCM sensor, metal oxide semiconductor sensor, and polymer composite-based sensor, which are referred to as chemical sensors, are less efficient regarding their trace-level detection. In addition, it can be less efficient due to the molecules present in the air, which are referred to as interference noise molecules. According to Aeloor and Patil (2017), the above two issues can be solved by integrating the system with other detection mechanisms like gas chromatography and by adding gas filters to screen noisy molecules, respectively.

**Table 10.1** Examples of sensor types and their properties (Karakaya, Ulucan, & Turkan, 2020).

| Sensor type | Detection range | Usage area | Advantages | Disadvantages |
|---|---|---|---|---|
| Metal oxide | 5–500 ppm | Food and beverage industry Indoor and outdoor monitoring | Suitable to a range of gases Operation in high temperature High power consumption Fast response, small size, easy to use | High sensitivity and specificity Sulfur poisoning Weak precision Humidity sensitive |
| Conducting polymer | 0.1–100 ppm | Medical industry Pharmaceutical industry Food and beverage industry. Environmental monitoring | Sensitive to a range of gases Fast response and low cost Resistant to sensor poisoning Use at room temperature | Humidity sensitive Temperature sensitive Limited sensor life Affected from drift |
| Quartz crystal microbalance | 1.5 Hz/ppm 1 ng mass change | Pharmaceutical industry Environmental monitoring Food industry Security systems | Good sensitivity Low detection limits Fast response | Hard to implement Poor signal-to-noise ratio Humidity sensitive Temperature sensitive |
| Acoustic wave | 100–400 MHz | Environmental monitoring Food and beverage industry Chemical detection Automotive industry | Small size and low cost Good sensitivity and response time Response to nearly all gases | Hard to implement Temperature sensitive Poor signal-to-noise ratio |
| Electrochemical | 0–1000 ppm adjustable | Security systems Industrial applications Medical applications | Power-efficient and robust High range operation temperature Sensitive to diverse gases | Large size Limited sensitivity |
| Catalytic bead | Large scale | Environmental monitoring Chemical monitoring | Fast response High specificity for combustible gases | Operates in high temperature Only for compounds with oxygen |

*(Continued)*

**Table 10.1** Continued

| Sensor type | Detection range | Usage area | Advantages | Disadvantages |
|---|---|---|---|---|
| Optical | Changes with · light parameters low ppb | Biomedical applications. Environmental monitoring | Low cost and light weight Immune to electromagnetic interference Rapid and very high sensitivity | Hard to implement Low portability Affected by light interference |

*Source*: Karakaya D., Ulucan O., Turkan M. (2020) Electronic nose and its applications: A survey international. *Journal of Automation and Computing* 17(2), 179–209. https://doi.org/10.1007/s11633-019-1212-9. Springer/. https://link.springer.com/content/pdf/10.1007/s11633-019-1212-9.pdf.

Noise removal or reduction in noise, which is generated because of the sensor signal possibly because of error during data analysis or from changes in odor strength, is referred to in literature as white noise. The noise can be reduced using one of the methods available, such as Fourier transforms or Kalman filters.

So how do we know what are the characteristics of a good sensor? According to Wilson and Baietto (2009), the following factors should be considered when selecting a sensor:

1. Highest sensitivity—in relation to the chemical compound(s) of the sample. The threshold of detection can be close to the human nose, that is, down to about $10^{-12}$ g/mL.
2. Can function with diverse detection capabilities.
3. Relatively has a lower selectivity in order to be sensitive to a wide range of different odors.
4. In open environment, such as field applications, the sensor selected should have lower sensitivity, including humidity and change of temperature.
5. The selected sensor should have shorter calibration and training requirements.
6. The capability to function at a relatively lower temperature, whenever the need arises.
7. Faster recovery time.
8. Shorter recording and analysis times.
9. Higher sensor or sensors stability.

According to Di Rosa, Leone, Cheli, and Chiofalo (2017), techniques to overcome sensor shortcomings may include:

● The integration of electronic nose analysis with gas chromatography.
● Hybrid electronic noses with combination of different sensor technologies, such as the use of MOS, MOSFET, and BAW in different combinations.

*Smart Sensor (SS)* SS is made up from a number of components, such as transducer (sensing element) amplification, signal-conditioning modules, A/D converter, memory, and logic control. The above can be described further as a primary sensing element, excitation control, amplification, analog filtering, data conversion, compensation, digital information processing, digital communications processing, and SS power supply.

# 10.5 Online processing and network

The MOD has been designed to provide and manage real-time data from the location where the device is functioning to any location around the world (a wireless network), for example, via Internet and mobile phone or from one outdoor location to the analysis station.

The MOD devices can be also connected in the form of a network if monitoring is needed over a wider field or area. For this purpose, several MOD devices will be located at different points. These devices will be sending various data to be analyzed simultaneously, on a 24-hour basis, if this is necessary. This kind of network can be very useful specifically in remote monitoring areas. The type of network to be selected depends largely on the system required by the user; however, any of the major types of network, such as local area network, wireless local area network (WLAN), metropolitan area network (MAN), or wide area network can be implemented. However, for the design in this book, a WLAN is used to the gateway. As the signal from the MODs will be received by the gateway, these sensors' outputs (signals) will be received by the main control station via, for example, Ethernet. Access of vital data and remote control of the system will save time and reduce cost over time.

# 10.6 Embedded MOD

Research and development around the world are presently taking place for the purpose of embedding MOD into various devices, including robotic machines. This is part of further effort in mimicking nature by adding to our devices a function similar to our sense of smell. Adding MOD to a variety of devices that may have one or more functions, other than the detection of odor, will enhance the capability of the device further and consequently will reduce costs, minimize wastes, and may provide better energy efficiency as well as reducing the number of working hours/labors.

## 10.6.1 The importance of embedding MOD

Bringing to your mind any device that you may think embedding an MOD within it will enhance its function! A simple example is where food is stored, such as fridge or freezer. The MOD will enable the monitoring of food quality. Embedding firefighters with some of the devices or equipment they use in detecting toxic gases. Lack of oxygen or other types of dangerous emanations underground such as during mining, embedded MOD will save lives. By the addition of MOD to various types of transportation systems, as part of the vehicle, will provide additional safety. MOD can be also embedded successfully in robotic machines and mobile robots. The above are just few examples out of hundreds or more ways in using embedded MOD as one of the vital parts of any system.

## 10.6.2 Robots (Mobile)

By understanding the behavior of animals, especially lower species, in the way detecting certain odor occur, the knowledge gained will provide us with ideas and methods on how to provide accurate and simplified system for mobile robots. A good example of species in understanding the ability in detecting certain chemicals are moths and ants. According to Hiroshi Ishida and Toyosaka Moriizumi (Chapter 16: Handbook of Machine Olfaction by Pearce, Schiffman, & Gardner, 2002) a male moth detecting a patch of a pheromone plume will follow certain behavior in order to stay within the path of the plume that has been released from a female moth:

> In contrast to the simple chemotactic behavior of bacteria, the fundamental behavioral strategy of moths is upwind flight (anemotaxis) triggered by olfactory cues. When a male moth encounters a patch of a pheromone plume, it turns and progresses in the upwind direction. As long as the moth is flying in the plume, repeated "upwind surges" bring the moth closer to the female. When the male accidentally leaves the plume and the pheromone signal is lost, it starts to fly from side to side across the wind with a gradually broadening scanning area. This behavior is called "casting," and is effective in relocating the lost plume.

In trying to understand the process in finding the origin of an odor, such as chemical plume, certain species can easily detect and follow its path, uniform or nonuniform flowing field. Embedding an MOD within a robot to behave in a similar way may not be possible without accelerating the response time within the device—sensitivity of the sensor—that is, as close as possible with those responses detected in animals, some of them can be in the order of 100 Ms (Arbas, Willis, & Kanzaki, 1993). Following a uniform or changing flow field may need comprehensive multiphase algorithm training, such as the use of ANN. Also, specifically designed sensors that can sense at high-speed movement would help a great deal in detecting the odor compared with the present design, which is mostly for stationary devices.

It is essential to train a mobile robot on how to locate an odor or a plume when it is not possible by humans to detect it. The training also should cover when the odor or the plume flow is located in another location other than where the robot is stationed, that is, a situation where the robot has to move and search around in order to locate an odor. As soon as the robot locates the highest concentration of an odor in a certain spot—which can be achieved by taking different air samples while moving within a particular area—meaning the robot located the source and, therefore terminates the search.

# 10.7 MOD market

Presently there are factors that are supporting and pushing positively forward the MOD market, mainly due to advancement in recent technologies such as AI, Cloud, IoT, and advancement in sensor development. These technologies presently form an important part of the overall function of any MOD. Present prices for MODs range between $5000 and $90,000 or more and have begun to fall recently, mainly due to the development of the printed electronic circuit and 3D printing approach, which meant that investors, manufacturers, and customers began to notice the beginning of a vast expanding market where the MOD can be used increasingly across variety of businesses, industries, governmental, and domestic use.

Competitiveness in the newly established MOD market has already a number of major players. However, the global market is still in the process of trying to establish itself, and therefore these major players are presently competing with each other for a bigger slice of the global market in the way they are expanding their customer base within variety of commercial, governmental, and various types of industry users that have not yet ventured fully the MOD market yet.

Market researchers in this field are pointing the finger toward North America and China as the biggest market for the MOD. According to ReportLinker, published in July, 2020, global MOD market to reach $39.6 million by 2027. For the year 2020 the electronic nose is estimated at US $19.9 million, according to the above source. While at the same time, it is projected to reach a revised size of US $39.6 million during 2027, that is, around 10.4% of compound annual growth rate (CAGR) over the period 2020–27. On the other hand, the MOS sensors will have a growth of 11.2% CAGR and it is estimated to reach US $18.9 million by 2027. In regard to QCM sensor, is estimated to grow by 9% CAGR during the next 7 years.

The above report also estimated individual market's growth such as the US market, it is estimated at $5.9 million with similar growth in the MOD market. However, higher growth is expected in China as it is estimated to grow by 9.6% CAGR and its projected market is estimated to reach a market size of US $6.8 million by the year 2027. On the other hand, Japan's estimated growth is 9.5% while Canada's growth is estimated to grow by 8.4% for both countries between 2020 and 2027. The biggest growth in Europe is taking place in Germany, as has been estimated it will grow by 8.3 CAGR by the same period indicated above.

There is higher global growth related to conducting polymer as estimated to be 10.5% CAGR. Most of this growth will originate from countries such as the United States, Canada, Japan, China, and Europe, taking the largest lion share by 10.6% CAGR. The total combined market for the above countries is US $4.4 million during the year 2020 and will reach US $8.9 million by the year 2027. For the countries within the market in Asia-Pacific, such as Australia, India, and South Korea, the forecast by the year 2027 is estimated at around US $4.6 million.

The same report cites examples of competitors within the above market, which are Alpha MOS France, Aryballe Technologies, Plasmion GmbH, Electronic Sensor Technology, Inc., Odotech Canada, AIRSENSE Analytical GmbH, and The eNose Company.

Finally, as the market witnesses increase of MOD usage within the medical field for disease screening, such as via exhaled breath testing, that is, as a noninvasive way in diagnosing the type of microbial infection that helps in saving time and cost, and more importantly, a quicker way to treat a patient. However, the speed of using MOD within the medical field will accelerate even further as and when wider applications are provided within the device, in addition to enhancing the reliability. For these reasons, the markets in various parts of the world are coming forward to participate in the form of investors, researchers, and vendors helping to speed up the development of the MOD and establish it as a vital product on the global market.

# 10.8 MOD future development

The MOD development will continue for the foreseeable future, specifically as the cost of technology improves and the prices associated with it will fall, in a similar way as to what happened to the mobile phone when it appeared in the market during the early 1980s. After a mere few years the prices started to fall and eventually a mobile phone that cost £1500 in the eighties, can be bought for few pounds at the present time. The market will witness a similar scenario when it comes to the future or even the near future for a price of one MOD!

In regard to how far the technological development may continue advancing in this field, this may depend on a number of factors, such as the appearance on the market of a new device that competes with the present devices but follows different approaches and/or technology and/or offers far more than what the present MOD offers, plus other factors such as materials and resources, market, changes methods in manufacturing and related industrial factors. However, for the MOD as it is now and as it will be in the near future, the vast predicted investments will push forward for further development of the device and will be a major player across the globe as an essential part of everyday life.

There are already approaches to combine different technologies designed for various elements and specialized odors detection within one device, that is, a hybrid system. For example, using a dual-column gas chromatograph with flame ionization detector in addition to MOD system. The hybrid device used, such as within a medical field, has the capabilities of breath-profiling plus the capabilities for chemical analysis, which means the outcome of the breath composition will enable doctors to identify the type of disease on the spot rather than having to send the patient to a medical laboratory or a hospital for further investigation. This kind of a hybrid device can work in various other situations, whether within an internal or open environment. Hybrid system is also used in the international space station, specifically to monitor the internal space station environment, in relation to the health and safety of the crew.

Hybrid system can be in the form of electronic nose (MOD) and electronic tongue or electronic nose, electronic tongue and computer vision systems, that is, nose, tongue, and eyes are different types of hybrid systems that have been studied commercially and via a number of researchers for more than a decade, as it is used within the food industry. The food industry requires fast and cost-effective approaches as a way to check food quality and safety. This kind of hybrid system, combining a number of artificial senses, has been termed *sensor fusion*.

# References

Aeloor, D., & Patil, N. (2017) A survey on odour detection sensors, *International Conference on Inventive Systems and Control (ICISC)* (pp. 1–5). Coimbatore. https://doi.org/10.1109/ICISC.2017.8068688.

Arbas, E. A., Willis, M. A., & Kanzaki, R. (1993). In R. D. Beer, R. E. Ritzmann, & T. McKenna (Eds.), *Biological neural networks in invertebrate neuroethology and robotics*. San Diego: Academic Press.

Bax, C., Sironi, S., & Capelli, L. (2020). How can odors be measured? An overview of methods and their applications. *Atmosphere, 11*, 92. Available from https://doi.org/10.3390/atmos11010092.

Berger M. (2007) Nanotechnology electronic noses. Nanowerk. <https://www.nanowerk.com/spotlight/spotid = 3331.php>.

Burlachenko J., Kruglenko I., Snopok B., Persaud K. (2016) Sample handling for electronic nose technology: State of the art and future trends. TrAC Trends in Analytical Chemistry. 82, pp 222–236. <https://www.sciencedirect.com/science/article/pii/S0165993616300279>.

Di Rosa, A., Leone, F., Cheli, F., & Chiofalo, V. (2017). Fusion of electronic nose, electronic tongue and computer vision for animal source food authentication and quality assessment — A review. *Journal of Food Engineering, 210*, 62–75. Available from https://doi.org/10.1016/j.jfoodeng.2017.04.024, ISSN 0260–8774.

Full, J., Delbrück, L., Sauer, A., & Miehe, R. (2020). Market perspectives and future fields of application of odor detection biosensors—A systematic analysis. *Proceedings, 60*, 40. Available from https://doi.org/10.3390/IECB2020-0702.

Hierlemann, A., Weimar, U., & Baltes, H. (2002). Hand-held and palm-top chemical microsensor systems for gas analysis. In T. Pearce, S. Schiffman, & J. Gardener (Eds.), *Handbook of machine olfaction*. Print ISBN:9783527303588 |Online ISBN:9783527601592|. https://doi.org/10.1002/3527601597.

Karakaya, D., Ulucan, O., & Turkan, M. (2020). Electronic nose and its applications: A survey international. *Journal of Automation and Computing, 17*(2), 179–209. Available from https://doi.org/10.1007/s11633-019-1212-9, <https://link.springer.com/content/pdf/10.1007/s11633-019-1212-9.pdf>.

Pearce, T., Schiffman, S., & Gardner, J. (2002) Handbook of machine olfaction. Print ISBN:9783527303588 |Online ISBN:9783527601592 https://doi.org/10.1002/3527601597, <https://onlinelibrary.wiley.com/doi/book/10.1002/3527601597>.

ReportLinker (2020) Global electronic nose industry. <https://www.reportlinker.com/p05819053/Global-Electronic-Nose-Industry.html?utm_source = GNW>.

Santos, G., Alves, C. P'adua, A. Gamboa, H., Palma, S., & Roque, A. (2019). An optimized e-nose for efficient volatile sensing and discrimination. *BIODEVICES 2019—12th International Conference on Biomedical Electronics and Devices.* <https://www.scitepress.org/Papers/2019/73907/73907.pdf>.

Silvello, G., & Alcarde, A. (2020). Experimental design and chemometric techniques applied in electronic nose analysis of wood-aged sugar cane spirit (*cachaça*). *Journal of Agriculture and Food Research, 2,* 100037. <https://www.sciencedirect.com/science/article/pii/S2666154320300181>.

Wilson, D. A., & Baietto, M. (2009). Applications and advances in electronic-nose technologies. *Sensors, 9,* 5099—5148. Available from https://doi.org/10.3390/s90705099. https://www.mdpi.com/1424-8220/9/7/5099/htm.

Zang, W., Liu, T., Ueland, M., Forbes, S., Wang, R., & Su, S. (2020). Design of an efficient electronic nose system for odour analysis and assessment. *Measurement, 165,* 108089. <https://reader.elsevier.com/reader/sd/pii/S0263224120306278?token = FB4372D82C67 5F3887A3F9FA17CC2F6138A361E1D32151B0642D73D5B1E5DAB9CC778B3377803B333 3F2EB0BD4B2A6EF>.

# Index

*Note*: Page numbers followed by "*f*", "*t*", and "*b*" refer to figures, tables, and boxes, respectively.

# Index

# Index

Printed in the United States
by Baker & Taylor Publisher Services